# Reefscape

*Reflections on the Great Barrier Reef*

Rosaleen
Love

JOSEPH HENRY PRESS
WASHINGTON, D.C.

**Joseph Henry Press • 2101 Constitution Avenue, N.W. • Washington, D.C. 20418**

The Joseph Henry Press, an imprint of the National Academy Press, was created with the goal of making books on science, technology, and health more widely available to professionals and the public. Joseph Henry was one of the founders of the National Academy of Sciences and a leader of early American science.



**Library of Congress Cataloging-in-Publication Data**

Love, Rosaleen.
Reefscape : reflections on the Great Barrier Reef / Rosaleen Love.
p. cm.
Includes bibliographical references (p. ).
ISBN 0-309-07260-3 (alk. paper)
1. Natural history—Australia—Great Barrier Reef (Qld.) 2. Great Barrier Reef (Qld.)—History. 3. Great Barrier Reef (Qld.)—Description and travel. I. Title.

QH197 .L68 2001
508.943—dc21

2001024281

The author and publishers would like to thank the following for use of copyrighted material: Lines from "The Beginning" (p. 211) by Mark O'Connor in *Firestick Farming: Selected Poems 1972-90*, Hale & Iremonger, 1990. Lines from Lao-Tzu's *Tao Te Ching* (pp. 66-67) by Ursula K. Le Guin © 1977. Reprinted by arrangement with Shambhala Publications, Inc., Boston, www.shambhala.com.

Cover design by Sandra Nobes.
Cover photographs by Peter Lik and Robert Halstead, IPL Image Group.



Published in Australia by Allen & Unwin Pty Ltd

Printed in the United States of America.

# About the Author

Rosaleen Love completed her Ph.D. in history and philosophy of science at the University of Melbourne and has been an academic as well as a writer and commentator on science and culture in the general media. Her books include an anthology of Australian science writing, *If Atoms Could Talk* (Greenhouse Press, 1987), and two collections of short fiction with the Women's Press, *The Total Devotion Machine* (1989) and *Evolution Annie* (1993). Her short stories and essays have been selected for inclusion in many anthologies in Australia, Britain, and the United States. Rosaleen Love's works spring from an abiding interest in the history of ideas—including wrong ideas—from science to futures studies. She is an invited member of the Humanity 3000 seminar series for 1999-2000, organized by the Foundation for the Future, Seattle.

# Contents

Torres Strait
PAPUA NEW GUINEA
Thursday Is.
Raine Is.
Cape Direction
Night Is.
Princess Charlotte Bay
Flinders Group
Lockhart Mission
Nesbit River
Nymph Is.
Lizard Is.
N
Chesterfield Islands
Cooktown
Michaelmas Cay
Green Is.
Port Douglas
Cairns
Hinchinbrook Is.
Cardwell
Great Palm Is.
Magnetic Is.
Townsville
Whitsunday Group
QUEENSLAND
Swain Reef
Yeppoon
Rockhampton
Gladstone
Heron Is.
Sandy P.
Fraser Is.
Brisbane
PAPUA NEW GUINEA
Gogo Station
AUSTRALIA
Great Barrier Reef
0 200k. 400k.

# Introduction

I settle into a large sandy hole previously excavated by a turtle. Above me the pale furry leaves of an argusia bush provide some small shade. The southeast trade wind blows softly from the Coral Sea. White sands shelve steeply down into clear blue-green water. A large white black-masked gannet and its equally large fluffy white chick peer from their makeshift nest of coral rubble on the sand at the high-water line. The salt air has a certain tang to it: half chicken-roost, half desiccated compost.

I turn to look back at the flat expanse of the island..I see thousands upon thousands of nesting sea birds: sooty terns. About me lies the sea with its horizon of islands and white waters breaking on distant reefs.

This island has grown from the sea, as corals grow to the light. Polyp by coral polyp, it has grown from the top of a sea mountain, now beneath the ocean and still subsiding. Beneath me the mountain sinks. The ring of breakers all around and within them the sea calm and flat, the waters clear, the sky a brilliant blue: I feel I have traveled so far that I've arrived at the shores of quite another world. I have sailed beyond the Great Barrier Reef to the Islands of the Blessed.

I am in French territory: Renard Island in the Chesterfield Islands some 500 miles northwest of New Caledonia.

I am happy.

✩

This is a book about the Great Barrier Reef of Australia, but I begin with the islands that lie beyond. Happiness is such a place, the place that lies beyond.

In August 1998 the Chesterfield Islands hit the headlines when they were visited by a man who found happiness of a certain kind, the happiness of being unexpectedly alive. He was Stephen Fossett, the British millionaire balloonist. He wanted to be the first person to fly around the world in a balloon, and the Chesterfield Islands were as far as he got. He flew into a severe storm, his balloon was wrecked, and he plummeted seaward from the sky. Money bought him his quirky variety of solo-ballooning happiness. Being wrecked near expert Australian rescue teams was an arbitrary good luck factor. This man survived the plunge of his balloon to newfound delight in life.

I sit in the shade of an argusia bush and reflect on the nature of happiness. This argusia bush is part of the advance guard of the greening of a coral cay. Birds stop by, bringing plant seeds in their plumage and their digestive tracts. Grasses and small hardy shrubs come first, trees later, when deeper soil has had a chance to get established. The argusia bush is hardy and resistant to the digging activities of nesting turtles. As I look out from my hole, I count eight other holes that turtles have scooped out at the base of this particular bush. Yet the bush lives on. As one branch dies, a new branch springs from the base. The bush radiates dead and living arms in all directions. Its flowers are like green baby calimari, hence the common name of "octopus bush." The dead arms fall to the ground and form a dense, protective mat.

Above my head in the argusia bush a red-footed booby chick, the size of a domestic fowl, peers down at me. It will not move just because I have temporarily claimed this turtle excavation for myself. The red-footed boobies are mostly to be found in this region of the world. They nest in the straggly trees. Boobies that nest on the ground, the masked gannet and the brown booby, are also found on remote cays of the Great Barrier Reef. The red-footed booby nests in trees as a survival strategy, but since there are so few trees, it has made a choice of ecological niche that limits other choices. World domination is out.

A shady hole, large enough to fit a human or three comfortably, is a good place from which to view this blessed place. The turtle that dug this hole will not be back to reclaim it. She will never see her young hatch some three months hence.

A small plastic bottle surfaces under my lightly digging fingers. So do small hopping creatures the size of fleas. Who knows what other microarthropods are lurking in the sand beneath, interstitial fauna inhabiting the chinks of the world. A small spider parachutes from a leaf at the top of the argusia bush onto the open pages of my notebook.

I am alone in this place of great beauty. Forget my companions, the 26 people who came with me on this expedition. They are not at this moment sharing this particular argusia bush. Forget the several hundred thousand birds screeching behind me as the day warms up. I am alone yet immersed in a web of fantastic life.

The water in front of me beckons with the green-blue milky-white opalescence peculiar to reef shallows. There is another world beneath the water, a reefscape of living rock and vibrant swimming life. Mask and snorkel offer temporary sojourn in this other place. I move from the world of air to the world of water and see from underneath how the ripples on the surface of the water mark the upper boundary of the new world and serve to diffuse the harsh tropical sunlight. Underneath, lumps of coral rubble roll this way and that on the sandy reef floor. Soon fishes will appear and then whatever else skitters and slithers and lurks below.

On a trip to the Great Barrier Reef early in 1999 I went into a dive shop to hire some dive gear. "Have a good holiday. Enjoy the reef while you can," said the shop assistant. I thought then that he meant "Enjoy it while you can," in the "life is fleeting" sense. Later I realized he meant "Enjoy it while it's still here." In 1998 a dramatic wave of coral bleaching spread across the tropical oceans of the world. Corals from most species bleached to white, suddenly, and within weeks. First noticed in December 1997 in the Galapagos Islands, the bleaching swept across the Pacific Ocean to the Great Barrier Reef and onward, ultimately affecting corals in the Caribbean some nine months later. The colors of corals bring pleasure; their widespread bleaching brought dismay. While some corals have recovered, others have since died. The situation is disturbing.

The Great Barrier Reef has existed in its present form for roughly 6,000 years, which is only a moment in geological time, but the reef's moment may be passing.

Everything flows, says the philosopher Heraclitus, everything moves on. All things are in a state of becoming and perishing. If Heraclitus, who lived in the Greek city of Ephesus in the fifth century B.C., had slipped underwater at the reef then and now, he would see some proof for his philosophy. The reef is not a once-

and-for-all-times creation but rather an organic system that exists in a state of dynamic tension. The reef waters into which I slide today are not the waters I entered yesterday. Change in the pattern and distribution of species is a fact of reef life, whether caused by cyclones, global warming, ice ages—or the fact that humans are for the moment the dominant species and are changing all that they touch. Catch the watery moment as it flows by.

Just as nineteenth-century artists painted Australian landscapes as if they were European, so the reefscapes of Oceania were first imagined as coral plant gardens in a quite literal sense. Engravings published in a book in 1755 show corals growing on empty oyster shells scattered over the sea floor. The shells serve the function of anchors, the underwater equivalent of roots, and from these firm bases the corals stand up straight and tall as fern fronds in a European conservatory.[1] Then they were *rock* gardens. Learning to see the reef as a living, feeding, breeding, organic community came much later.

Today the Great Barrier Reef is renowned as one of the natural wonders of the world, recognized by UNESCO in its listing as a World Heritage area of outstanding universal value, important both for its natural beauty and its cultural value to Aboriginal people. Astronauts see it from space; it is the largest structure in the world created by living things—not humans but microscopic creatures, the coral polyps. Millions of other weird and wild forms of life live on the reef, or inside it, or around it. Some of them are people. Travelers come to the reef in search of a contrast to their usual lives. The reef offers beauty, wildness, and rest for the weary citified soul. It wasn't always perceived this way.

It is not so much the reef as it exists in and of itself that I want to explore in the pages that follow, though it will of course be part of the story. It is more the *meaning* of the Great Barrier Reef as it is encountered in danger, work, fun, or the search for sustenance. It is a place of possible and impossible dreams. Reefscape is more than underwater landscape. Reefscape is where the rocks are living, plants grow downward, the sky glows through a refracting barrier of water, and there is no air save that which humans take with them. On land, wildlife flees the human encounter; underwater, fishes move largely indifferent to human intrusion. Reefscape provides quite a different encounter with wildness.

One aspect of delight in nature is its unself-consciousness in a form of disembodied pleasure. Delight in nature takes one out of the pose of self-awareness: that this is me seeing and feeling this object. The awareness of being aware dissolves in delight. The everyday distance closes—that distance between the person who looks out at the world and the world itself. Viewing underwater increases the displacement of self. People must brave the transition between the element of air and the element of water, the moment when two different worlds join. That view, half in and half out of water, is a moment where one is here but not yet here, there but not yet there. Catch me if you can. I am suspended in warm water as if I were a part of it, entering into some kind of prehuman condition of flowingness. The history of human evolution rolls backward for this moment of reentry into the oceanic past, and the planktonic ego is liberated, for the moment, free from responsibility for the earth. Air and sky are left behind, the sea washes over, and it is a different self that dives down to meet whatever moves below.

The Great Barrier Reef is listed as a World Heritage area for its cultural as well as its natural sites. Coastal places have stories to tell for both Aboriginal and non-Aboriginal people. The World Heritage area includes Aboriginal middens and sacred sites as well as historical shipwrecks and ruins of early European settlement. The sea and the coastal lands provided rich sustenance for Aboriginal people for the past several thousand years. In so many different ways, reefs have become constructs of human imagination, hopes, and fears. Sea places have stories to tell, of how other worlds connect to this world, of how this place came into being. The stories of coastal Aboriginal people are tales of sea creatures and their journeys, stories that connect past mythic events with present coastal land and reefscapes. Here a spirit ancestor chased a whale or a dugong; there it lay down to rest—and if you look with the eyes of faith, if you see its shape in the rock and its breath in the spray of the waves, the reef shimmers with mythic significance.

If I make a story about the reef I may start with a certain place and time, perhaps most obviously the reef's origins in the growth of coral polyps to the light, their death and consolidation. There is a specific past to this biological and geological structure and a future that is foreshadowed. The geological aspect spans the long course of global reef construction and extinction—so far nine periods of mass extinctions followed by reef renewals. Add in the historical aspect of the reef, which brings into focus the reef as resource—a story of commodities and extractive industries, from fishing to mining. These were years of destructive culture contact, as Indigenous people met strangers from over the seas, with their guns and hunger for low-wage labor, their development plans that paid no heed to the Indigenous peoples' rights to the land and sea.

The reef has seen shelling and sheep farming, pearl and *bêche de mer* fisheries, a virtual penal colony on Palm Island, a leper colony on Fantome Island. Limestone was mined; drugs and people were smuggled. Events of the relatively recent past merge into the present growth of million-dollar tourist industries, where reefscape gains new meanings of leisure and pleasure, and reef managers gain new jobs protecting marine parks and World Heritage.

I set out in search of reef experience and meaning and spirit, wanting it all. Going out and seeing for yourself gets you only so far; but reaching into reefscape through the imagination has few limits. Even traveling in time is not a problem for the armchair traveler. Nor is traveling into the microworld of the reef. Film and video increasingly allow ever more impossible journeys into unimaginable places, like the interior of the coral polyp even as it spawns and tiny new forms of coral life pulse their way into the plankton. I know I shall never travel into the underwater caves in Palau, where turtles get trapped and the sandy cave floor is littered with their bones, nor swim underwater with crocodiles. I won't travel there myself, but I can still experience it from the comfort of the couch.

Part of the emotional reaction to the reef is a feeling of freedom, for a time, from the chaos of human affairs. This can mean getting away from friends and lovers, getting away from human relationships with all their unspoken rules and complications. Nature's chaos liberates as it energizes. Travel as escape to freedom is part of the story. So is the search for the part of yourself that you haven't found yet but that you know is there, within the workaday self, waiting to be unleashed: the snorkeller bursts forth from the couch potato; the diver lurks within the public transport commuter. Travel to the reef involves a double displacement of the self. The traveler goes out, and the traveler goes under.

Knowledge is a motivator, too. Knowing yourself is one small part of it. There is so much other knowledge out there waiting to be wrestled into some kind of order. Science on the reef embraces so many diverse and alluring fields: marine ecology, oceanography, plate tectonics, ethnobiology, and more. I come as a tourist to reef sciences, gleaning a thread of knowledge from here, weird and wonderful facts from everywhere—and some of them are true. I also come as a philosophical traveler, hoping to understand something of the human response to this exotic other world of nature.

I'm aware that I will be able to know only a tiny fraction of reefscape. The human intellectual response to this world, at least until now, has been limited by the capacity of the human brain. Imagine, though, a future in which robot turtles accompany real turtles on their migrations, sending back all possible information to computer expert systems. Robot turtles will collate information from reef sciences and industrial waste processes and add to it knowledge of international maritime law and global finance. In the face of future robotics expertise, mine is a relatively laid-back enthusiasm. I want the large and wonderful ideas but not the level of detail with which the scientist, human, or robot must necessarily work. I want to know—at a grand, overarching level of knowing. I want to add in the personal dimension something the robot cannot yet accomplish. I want to know something about myself in relation to reefscape.

Whatever it is that I search for, I suspect I shall not find revelation of my unique place in the cosmos. The reef tells me to forget all that: I'm nothing special, just one among a million species that lives and eats and breeds and dies. Yet knowing these limitations, I still yearn to escape them. Having only one brain, one life, how limiting it is! Next life I want to be a nature photographer, or a reef scientist, or a prophetic visionary who preaches the gospel of wild places, or a marine archeologist on a wreck dive.

When I ask people why they have come to the reef—including reef scientists—they always say, "It's beautiful." I want to examine the nature of this particular delight, how it has happened that people know the reef as beautiful. Ask a recreational diver what it was like after a dive and you'll get a happy dippy grin. "Great," he'll say, she'll say. Then realizing more is expected, there might follow a comment (say of a night dive): "It's all those small red eyes that peer out at you, and after a bit you realize what they are. They're shrimp!" The diver's eyes are shining; the face is flushed to the pressure mark of the mask on forehead. Words are poor things in those first few moments back on land. The French philosopher Roland Barthes wrote: "Pleasure can be expressed in words, bliss cannot."[2] This is the sum of it, though Barthes was never thinking of bliss through night encounters with shrimps.

The recreational diver often likes to keep note of dives taken, where and when. The logbook is a record of adventure and a diary of underwater experiences. But divers grope with words to express the gap between the experience and the recollection. The diver wants to latch onto and hold the feeling of being there. This helps explain the popularity of underwater photography. To me, holding camera gear gets in the way of actually "being there," but, to many, film conveys better than words the immediacy of the underwater experience. Still, however technically brilliant it becomes, film cannot, by its nature, capture the elusive merging of the self into the oceanic moment. The reef has its own ambience, as do the forest and the mountain. The aspect of being there seems more to the point, when one *is* there, than capturing words or images to record the experience later. The adventurer bursts forth from the inner self to enjoin the narrator: Feel the flow! Catch the ambience! Enjoy! There is more to the reefscape experience than reading the surface and depths of the reef as a series of stories.

Going underwater is like going into another universe, as if going beyond the here and now of individual existence to something entirely other. It is not surprising to find that many people on dive boats read books on spirituality. Into the gap between the experience and the story to tell it, with the senses infused to near-total engagement, into this place of "losing oneself," the sense of empathic communion with watery nature approaches the territory of the otherworldly, the spiritual. Just such a spiritual journey was described by Augustine of Hippo, a bishop of the early Christian church who lived in North Africa between 354 and 430 A.D. From time to time Augustine found himself visited by a feeling of delight in nature, swept up in a flood of feeling and emotion. He came to welcome this visitation as something that mattered more to him than his earlier desire to achieve an impassive serenity of soul. Delight arrived unannounced, disturbing in its unanticipated pleasures. Augustine wrote of his pleasure in open fields and the sky and the Mediterranean sea: "There is the grandeur of the spectacle of the sea itself, as it slips on and off its many colors like robes, and now is all shades of green, now purple, now sky-blue."[3] The natural world was alive with delight and desire, and a yearning "that makes the heart deep" led him to reflect on the absent perfection of God here imperfectly glimpsed. He sought to know more of whatever it is that might lie beyond what is said or what might lie beyond what is presently knowable. The ambience of the natural world was one of the paths to this different way of knowing.

Yearning for a country that is forever distant, the reef voyager faces a dilemma in delight. Yearning is an example of what psychologists today dryly call a "complex emotion." Yearning has an element of despair in its hope. The object of devotion is never wholly attainable, at least here and now. The impossibility of the quest is an element of the quest. It is part of the experience of travel to be forever moving on, restless in the search for new wonders. The reef experience is transitory. The distant beloved is ever distant. Perfection is something that is present in the idea, never quite attainable in the experience. Even when we are in its midst, the underwater realm is always temporary. The moment of return to air and the imperfections of everyday life inevitably await the voyager.

Being in nature I feel so alive yet am reminded of my own mortality, so aware of myself yet also of my insignificance in the flow of creation and destruction. Instead of following physicists who find God in the new physics, I wanted to explore why the reef equally inspires a sense of awe. Instead of the Tao of physics, I'd go with the flow of reef waters. Where planets react passively to cosmic forces, reefs, in self-actualizing fashion, create their own dynamics. People, in turn, create meaning from nature. From contemplation of both the very large and the very small comes a strong sense of human finitude.

Reefscape is a place where Indigenous people situate spiritual power. The stories of coastal Aboriginal people invest reef places with a spiritual significance that finds expression in myth and ceremony. As a traveler to the reef, I glimpsed these stories, however imperfectly, when I walked the trail on Stanley Island in Princess Charlotte Bay in far north Queensland. I learned a little of how others have seen this place and their ways of communion with an interconnected web of life, differently imagined. The ways of science are different but equally capable of supporting a sense of something special about nature; there can be some kind of sacred dimension to science. I wanted to explore this notion and see where it got me.

The intense feeling of belonging to the broader community of life has long been part of many of the world's religious traditions. Now it is beginning to enter the ecological-scientific mainstream. When the Greek philosopher Plotinus described his experience of natural beauty, it was with "amazement and a shock of delight and wonder and passion and a happy excitement."[4] He glimpsed the connectedness of all sentient forms of life, a vision that today takes scientific form in ecology. A coral reef proves a living, pulsating example of connectedness.

The reef provides the vision of beauty; reef science supplies the theoretical links. Divers can see for themselves the interdependencies, once they know what to look for. For instance, at the fish-cleaning stations, cleaner fishes nibble the parasites from the gills and mouths of fishes much larger than they are. The anemone fish shelters within the waving fronds of the anemone host, enticing other small fishes into the anemone's trap. I see utterly different creatures—prickly sea urchins, sausagelike *bêche de mer*, sea stars, brittle stars, and feather stars—and find science has discovered their close relationship within the phylum of the Echinoderms. I know that all the crannies in the corals, within corals, and underneath coral rubble are occupied by forms of life, because the photographer-naturalist has shown me in advance. Fishes glide by languidly, and their indifference to human presence is our pleasure.

This conjunction of happiness in relation to ecological awareness is unusual. Usually it doesn't work this way. Nature writers lament the loss of natural beauty through human destructiveness. Ecological awareness tends to bring about a general gloominess of outlook. It's hard to remain happy with Greenpeace predicting that coral bleaching might well destroy the Great Barrier Reef within 20 years. It seems to be part of the human condition, this tendency to care about people and places and things that can and almost certainly will be lost. Yet the optimist, though sharing the pessimist's knowledge of ultimate loss, may still put ecological gloom to one side. The British philosopher Margaret Miles suggests that, to be happy, now and here the trick is to reflect on "the broader generosity of the universe, the continuous, amazing circulation of gifts."[5] Being part of this oceanic nature privileges us to share the fullness of its beauty and bounty. The sense of relatedness and interdependence of all life emerges from the effacement of the self.

Margaret Miles was talking more generally about her reading of ancient philosophers, but the Chesterfield Islands or the Great Barrier Reef would have illustrated her point perfectly. The broader generosity of the universe is nowhere more manifest than in the sight of hundreds of thousands of nesting birds or in gently sifting sand through your fingers and knowing it is composed of billions of skeletons of the tiny algae, the forams, that sustain the reef. This is what makes divers happy, glimpsing their corner of interconnectedness, knowing it flows indifferent to their presence and will continue to flow if they are not there.

Call it what you will—happiness, delight, rapture, enchantment. Romance, perhaps, captures something of the multiple aspects of yearning that the reef conjures: this faraway place, alien yet gorgeous, a back-to-the-womb suspension in warm, nurturing waters; the prize of access to it, and a kind of possession of it, in mind and spirit. The time of the traveler is "time out" from ordinary life, "time out" from work time. Work time is the time of the clock, minutes and hours becoming internalized in the human psyche. On the reef the traveler enters another kind of time: subjective time. Happiness is the "hereness" of the present, with no thought of later or somewhere else.

Of course, the diver must watch the flow of time as measured by the air gauge to check how much air is left in the tank. When the arrow hits the red mark at 50 bars of pressure, it's time for up and out. Anxiety about one's air supply is perfectly healthy, if a somewhat overwhelming concern of the novice. It is an anxiety that protects. Even the expert diver must guard against the "raptures of the deep"—staying too long, at too great a depth and thus risking the often fatal effects of nitrogen narcosis. The raptures of the deep may lead to increasing carelessness and death.

In 1998 Eileen and Tom Lonergan, a young American couple, were lost at sea after a dive on the outer edge of the Great Barrier Reef. They liked taking long, slow, and relatively shallow dives, dives in which they used their air supply economically and stayed under for as long as they could. On their last dive they rose to the surface afterward to find the dive boat had departed without them. A slate with a message for help signed by both of them showed they had survived the first night together in the open sea. It is terrible to imagine what it must have been like for them, to survive one day, one night, two days, however long it took, knowing their helplessness, knowing their vulnerability. Some of their dive gear washed up later on the coast, but they were never found.

The story of the Lonergans joins the story of baby Azaria Chamberlain as two Australian tragedies that have deeply affected many people, linking as they do the frailty of human lives with the fierce forces of nature. In August 1980 baby Azaria Chamberlain disappeared from a campsite at Uluru in the central Australian desert, a place of significance for many Australians. She was reportedly taken by a dingo, and her body was never found. Two years later her mother, Lindy Chamberlain, was convicted and imprisoned for killing her baby. Her conviction was quashed in 1988 when new evidence in support of her story was discovered.

What links the Lonergans' tragedy with the Chamberlain case is the wild mix of stories that sprang up, so many of them improbable and contradictory, in some kind of weird denial of the tragic fragility of life. According to some of these stories, the Lonergans faked their own disappearance, for insurance or other reasons. Lindy Chamberlain went to jail for six years because people did not believe her awful story. Terrible deaths in places of great beauty are the flip side of wildlife experience—the dredging from the depths, the primeval horror of being engulfed by the wild. No one seeks to become prey for the predator. Nature is exciting, but nature isn't nice. She has her glittering gardens through which monsters stalk.

I had two quite different introductions to the Great Barrier Reef. I first went many years ago to the resort at Heron Island on the southern edge of the reef, off the coast at Gladstone. Heron Island was then run by P&O Cruises, and there was a luxurious feeling of being shipwrecked in grand style. When I first walked off the coral sand and looked beneath the waters I felt delight imploding.

My second introduction to the reef was less conventional. When I came back, as I knew I must, it was to sail the Swain Reefs near Gladstone in the company of large plastic buckets of sea snakes. The snakes were temporary residents on the decks of the *Australiana*, skippered by Max Allen of Gladstone. Every few years scientists from the University of New England chartered the ship for 10 days of scientific research, and to make up the numbers they took on board-paying extras. My sister Kathleen, an Armidale zoologist, went along, and next time she took me. I am forever grateful to her for my sea snake introduction to the reef. Hal Heatwole was the leader of that expedition and its resident sea snake expert.

Each day young enthusiastic students climbed into dinghies and went out to snake-infested reefs. Leaping into the water, they held fast to their snake collection bags and, using only mask and snorkel, chased and caught the sea snakes, taking care to avoid any writhing, courting snake couples that tend to get tetchy if caught in the act. The snakes were brought back to the dinghies and emptied into plastic rubbish bins. The lid of each bin had to be raised and lowered fast as new snakes were added, as otherwise the snakes inside would poke their heads up and out and slither over. Sea snakes are highly venomous. Collecting venom was part of the reason for the research. Afterward, the sea snakes were returned to the sea, surprised but still very much alive and slithering. From Hal Heatwole I learned a fact that has never left me. The male sea snake has a double set of reproductive organs, covered in spines.

One of the delights of my reef life is a place called Nymph Island, which lies off the coast of Cooktown near Lizard Island. On navigation charts the island shows a peculiar amoeboid shape, with a huge hole in the middle as if the amoeba were about to divide in two. Coral islands are usually a fairly standard shape, roughly circular or elliptical, curving this way or that on the surrounding reef according to the direction of the prevailing winds. Nymph Island is different.

The beach rock on the shoreline is a conglomerate of shell and corals cemented together in intricate patterns. Near the mangroves a shallow stream flows swiftly out to sea. The shoreline changes from coral rubble to mangrove mud and sand. We wade up the stream into the interior of the island.

Imagine this: stories of dark interiors of dark continents, heart of darkness, heart of horror. Now turn the mental print to negative: imagine heart of radiance, heart of light. The swiftly flowing stream widens to a huge interior lagoon that spreads to remote edges lost in mangroves. Imagine: myths of inland Australia; tales of the vast inland sea; the search for the source of the Nile. Nymph Island lagoon has their aura of myth and mystery. Small stingrays dash past in the shallows, the tips of their "wings" skimming in air. The soft burbling of Torres Strait pigeons comes from the distant mangroves. A sea eagle's nest, a large untidy pile of sticks, sits perched on top of a half-dead tree.

This island is a miracle of recent times. Between 6,000 and 20,000 years ago, the seas rose, the land below subsided, and corals grew up to the light from a submerged hill along the former coastal plain. Corals grow and are broken and die, and coral rubble swirls round and accumulates. New corals grow on the debris of the old. Larvae of clams and sponges drift by and join with algae in helping create a reef. Nymph Island has grown from ocean, with the forces of wind and water and life. Sand piles up to reach the surface. Birds drop by and bring seeds in their feathers and digestive tracts. This island under my feet has come from an expanse of ocean, rubble, and seeds dropped casually by birds. I am walking on young ground.[6]

Coral rubble is heaped up all along the perimeter of the island, whole beaches of it mixed with shells and fragments of beach rock and ocean debris of all kinds—claws of crabs, bones of dead birds, odds and ends of wood, even curly blue wire. I pick up a piece of coral with delicate branches, the *Acropora* or staghorn coral. This dirty gray piece of coral still serves the moment of delight. If I were to look at photographs of the deep structure of this piece, I'd see down into the level of the skeleton of each coral polyp, the corallite. Each corallite appears white and delicately patterned with a radial symmetry like the hubcap of a fancy car or a circular snowflake. If I should look further at the ridges and walls of the corallite with the aid of an electron microscope, I'd find structures finer and finer still. If I were an expert, it is these fine structures that would help me define the species. There is a sense here that if I could keep looking, if the human eye could be assisted by yet-to-be-invented technologies, I would discover even more minute structures.

In this piece of coral in my hands, repeated over and over, is the life sequence of one coral polyp: from free-floating larva to polyp that settled, cloned itself, grew, reproduced, and died to become one corallite among many. One day it will be sand. It marks the end of a long line of descent from the beginnings of the *Acropora* genus some 40 million years ago. In the warm waters of ancient seas, one sturdy branch began to take shape from a base in one polyp that was slightly different from the rest. With time the *Acropora* branches became slender and more delicate, the branching patterns more complex. *Acropora* species in different environments diversified and flourished to create the 110 or so species that survive today.[7]

Sunset on the reef, as well as sunrise, when the water is still and calm, is a time of special delight. In the stillness, pink sea blends into pink air. The horizon vanishes; sea becomes as sky, the sky as sea. Clouds become hills. The ship sails upside down, like a plane doing stunts, and the world is inverted, and this new inversion of the world is the new order of things. There is no fixed point of reference around which the rest of the world is swinging. Only when night falls or sun rises is the spell broken. The ship sails right way up again and returns me to the world I know.

From luxury liner, albeit high on land, to boatload of boisterous zoologists, I have led a fortunate reef life. This book is a tour around the idea of the Great Barrier Reef as much as its constituent parts. I decided I wanted to explore the world of the reef, not only its facts but also its power to keep drawing me back and into it. Some of the book was written in the depths of the 1999 Melbourne winter, punctuated with news reports of widespread coral bleaching and the possible end of the Great Barrier Reef. Wrapped in a blanket, I sat in the glow of the computer screen, dog at my feet, and conjured up memories of water the color of heaven, sand the color of angel cake, and the distinctive fertilizer smell of the true coral cay.

# Diving for Oldies

This was going to be the year of learning to scuba dive. The realization came to me slowly. I was in that weird condition called life, which entails getting older year by year, and I was well into my fifty-eighth. My first thought was—no, diving has to be one of those things that have passed me by, like becoming a diva or a dancer. I'd been on boats where people went diving, and I'd watched the rigmarole of getting kitted up in diving gear. I could never do that, I'd tell myself. The gear is so heavy. Women are more buoyant than men—I am, that's for sure—women have to put on more weights around their waist. Would I be able to stand up? And that heavy air tank? I'd fall over with that on. Backwards. Maybe once under the water I'd be all right. Between land and water, though, that would be the tricky part. There seemed to be so many things in the way of it.

Yet I knew diving would be so much more than snorkeling and well worth a try. I'd once gone down in what's called a resort dive, where the uncertified diver goes down with a personal dive master, one to one. That was diving made easy. I walked into the water off the beach, and my dive master helped me every moment. She carried the heavy stuff. She fixed everything that needed fixing. She held my hand the whole time we were underwater. It was lovely, not only to go diving but to be nannied and coddled for 20 minutes. That was the soft-option dive. I loved it. I thought that if I did learn properly there would be too much responsibility involved. Scuba diving is always done in pairs, even on the training course. I couldn't falter because my diving buddy would be depending on me.

Another good reason for not learning to dive was that in Melbourne the water is Antarctic icy even in the middle of summer. I convinced myself that diving was out of the question. In April I went to Townsville on Australia's northeast coast, to James Cook University, where I spent a happy month of reading and research. All around me students were heading off for dive weekends. It's all very well for them, I told myself with the grumpiness of middle age; they're fit and 19 years old.

It was Joanne Hugues, a young French science student, who helped me over the next hurdle. "Why not learn to dive here in Townsville?" Of course! She'd been smart. She hadn't done her open-water course in the English Channel; she waited until she got to warm Queensland waters. There went one reason for not learning to dive.

I went back to Melbourne and shivered through the winter, and then I came back to Townsville. I was going to give it a go.

I was not sure I'd pass the required medical tests. Not because of any particular medical condition, just the general wear and tear that comes with getting on in life. I was near enough to 60, and 60—well, my father died suddenly at 61, and I had only a year or two to go to catch up with him. The year of learning to dive was the year of *memento mori*, "remember you must die"—yes, I know, but not before I learn this one last thing.

I rang around to find a doctor in Melbourne who was qualified to issue the requisite medical certificate for diving. A woman doctor took me through some questions about my health, the kind that might disallow me from the start. Did I have asthma or heart problems? And so on. When at the end of the questions I was still looking good, I broached the question of age rather timidly. There was silence at the other end of the phone for a moment, and I thought, hey, there must be a real problem here. Then in a clear authoritative doctor's voice, she said: "Well, I'm 53, and I've been diving for over 30 years. I just passed a woman last week as fit for diving, and she was 74. The oldest woman diving in the world today is 95 years old."

That fixed me. Who was I to fuss? I made the appointment and I passed the medical. I knew that if I failed I would be letting down the cause of 95-year-old women scuba divers. The test for balance nearly got me, but I clung onto the invisible line on the floor, eyes shut, one foot placed in front of the other, imaginary glue under each sock, wobbling but maintaining a more or less upright stance.

Later I discovered the identity of the woman who went diving at the age of 95. She is no ordinary oldie. Leni Riefenstahl was famous in her youth as an athlete and infamous as the director of *Olympia*, the film of the 1936 Berlin Olympic Games, and *Triumph of the Will*—two films of Nazi propaganda she made in the 1930s. Both films are regarded as brilliant in their cinematography while fatally flawed in their glowing evocation of Nazi ideology. When she learned to dive, she enrolled in the course under an assumed name and did not tell the truth about her age. She was then 73 but said she was 52. Her German dive instructor totally disapproved: too old at 52, he said. After her last dive trip in 1998 to Milne Bay in Papua New Guinea, she booked the next trip for 1999.

I found a fellow traveler, if a somewhat ghostly companion who would probably be appalled by my presumption. Jean Devanny was a writer who loved north Queensland and the reef. She was born in 1894 in New Zealand and died in 1962 in Townsville. Active in the Australian Communist Party for 20 years—from 1930 until she left it in 1950 (with a torrid period of expulsion in the middle)—she suffered and indeed was made to suffer for her political commitment. Yet as her daughter, Patricia Hurd, said, her mother was many things: "a pioneer in the women's liberation movement, as well as organizer and agitator for militant causes in general, an established writer, lecturer, naturalist, musician, anthropologist."[1] However, among all these, it is Jean Devanny the *diver* that I chose to take on my travels. I felt definitely spooked when I discovered that Jean Devanny was roughly my age when she first went diving. In her case, back in 1950, she had to gear up in the pearl sheller's lead corselet and copper helmet. A team of helpers on the surface worked the pumps to bring air down a tube to her helmet. She sank under the water carrying 75 pounds of lead about her body. That takes determination. She'll do for me as a diving role model.

Love of nature runs through much of Devanny's writing, yet this love was something she kept firmly in a separate compartment of her life from her commitment to political struggle. Karl Marx and Friedrich Engels were never nature writers. They never visited far north Queensland and the reef. If they had, they'd have gone fishing, and Marxism would have taken a different slant. Only recently have political thinkers such as Australian philosopher Arran Gare foregrounded nature in political philosophy. Jean Devanny kept her two loves separate: her membership in the Communist Party in one box, her work for the North Queensland Field Naturalists' Club in another. In her interests in both politics and environment, Jean Devanny would have found a more constructive use for her energy and enthusiasm today in the environmental movement.

The middle years of the twentieth century were a time when the riches of the sea seemed inexhaustible. Jean Devanny visited the reef well before the pressures of international tourism. Her love of the reef is couched in the language of the nature study and science of her time. Her politics enter only implicitly, when she describes the labor of reef workers, the mackerel fishermen, the trochus shellers, the laborers in the sugar cane farms of the tropical coast. She valued their work for its part in the international struggle of all workers to gain the just rewards of their labor. Marxist theory did not go into the question of the limits nature might set to the growth of extractive reef industries, nor the effects of coastal agriculture and its fertilizer runoff to the reef, nor in the earlier years of this century did science.

Jean Devanny ultimately found little joy in political struggle. She had commitment, energy, activism, and comradeship (for a time) but rarely happiness. Yet happiness shines through in her nature writing, and what often kept her going in tough times was the simple pleasure of growing cabbages.

When Jean Devanny first went diving at Green Island, off the coast of Cairns, in 1950, the aqualung was not available to her. The invention of the regulator, a demand-intake valve drawing air from a portable cylinder of compressed air "on demand" as the diver inhales, is usually attributed to the French, to Jacques-Yves Cousteau and Emile Gagnan in 1942 (though its origins can be traced back to the early years of the nineteenth century. Swim fins were developed in the 1930s, and the single-lens facemask came later in the 1940s.) Naturalist and underwater filmmaker Noel Monkman claimed to be the first to import the Cousteau aqualung to Australia in the early 1950s. When in 1955 British science fiction writer Arthur C. Clarke and his companion Mike Wilson spent six months diving the length of the Great Barrier Reef, they reported that the locals were very curious about their diving gear. Their aqualungs must have been unusual then in Northern Australia.

Jean Devanny had long been a devotee of the waterscope, a simple wooden box with a glass plate on the bottom. Put it on top of the water and look through it, and the underwater world is made visible as the waves are flattened out. I remember my father making one by knocking the bottom out of a large empty Milo (a chocolate drink like Nesquik) tin and fitting glass across instead. We didn't have masks and snorkels: we had recycled Milo tins. But at least we could see underwater. Museum pieces, they'd be now, if anyone bothered to keep them.

Devanny's move from waterscope to diving was made possible through her friendship with naturalist-photographer Bruce Cummings and his wife, Frances. With the ingenuity of an ex-World War II soldier, Bruce Cummings built his own diving bell. It was a heavy steel cylinder large enough to fit one person and a movie camera. The camera fitted into an arm projecting out to one side. There were four large windows. Half a ton of lead ballast took the cylinder underwater. Air came from a pump manned by a tender on the jetty above (or the lugger, if it went out to sea). An air pipe was connected to the top of the cylinder. Another exhaust pipe brought used air back to the surface so that bubbles would not appear in the film.

One day back in 1950 or thereabouts, Jean Devanny entered the cylinder. The lid was screwed down. "An attack of nervousness seized me," she wrote. "I felt acutely alone, but as I was lowered away the amazing nature of the new experience effaced all consciousness of self." She was taken down 25 feet to the sandy ocean floor at the end of Green Island jetty. From the cramped quarters of the diving bell she looked out for the first time at underwater life, viewed from the bottom up. She raved about the jellyfishes, the corals encrusting the piles of the jetty, the parrot fishes, the schools of sardines, the "submarine volcanoes" of piled-up whirls of sand excavated by marine worms.

Then a message came down the two-way telephone from the jetty above: "You're going to have some visitors." The crew from a New Guinea riverboat free-dived down to greet her. When she published her account of this diving adventure in *The Australian Journal* she may have been pleased to hear her story had been chosen as the main feature. I wonder what she felt when she saw the cover illustration. She is depicted as a young blue-eyed blonde staring out of the diving bell in considerable alarm at three Papuan divers.[3] Jean Devanny would have been 56 years old at the time and had dark brown hair. She was not at all terrified of Papuan divers; she greatly respected their work and included their stories in her books. She was always fierce in the fight against racism. She was paid £20 for the article, an adequate price for the words but too little for the indignity.

For her next trip underwater Jean Devanny donned diving gear—but not as we know it. As she tells the story:

> "If you are game to go below in corselet and helmet," said Bruce to me one day, "I'll rig up the gear and you can get a close-up of the corals feeding in those patches off the end of the jetty. One of those gardens there is as good as you'll get anywhere."
>
> So on the next fine morning I sat on the railing on the jetty and with quickened heartbeat belying a fine show of

> bravado, waited while Bruce, Jack and Peter brought down the engine, weights, helmet, corselet, airlines and communication ropes and made the necessary preparations.[4]

In this gear, the half suit of the pearl divers—helmet and corselet only, no suit—the helmet simply rests on the diver's shoulders, held there by its own weight. The pressure of air coming down the hose from above keeps water from rising up the helmet. Jean Devanny was loaded with 75 pounds of lead weights. If she wanted to clear the window of fog she had to bend forward to let water in up to her chin, take a gulp of water, and spray it over the glass. Every snorkeller knows how spit on the mask does wonders for clearing fog, but we have it much easier than she did.

Devanny was a very scared woman that first time, she wrote of herself. But she went ahead with it. The lead was placed on her head and shoulders while she was chest deep in the water. She climbed farther down the ladder on the jetty until she dropped off the end into deep water. As she stood on the white coral sandy floor of the ocean, she was exhilarated. "I felt at home," she said. One step forward brought her nose smack up against the glass. As the current grew stronger she had to bend forward and dig her toes into the sand. But she loved it. "My spirits rose continually," she wrote. "This was the life! I felt native to the ocean. The feel of the water on the nude parts of my body was like silk." She managed to stay under until the weight of the helmet made her feel giddy. When she got out, she'd thought she'd been down five minutes, but 20 minutes had passed—about as long as a beginner's first scuba dive. And, as often with beginners, she'd forgotten something. In her case it was the communication cord on which she was supposed to pull to indicate distress. Luckily she didn't need it. After that first taste of diving, she went again, as often as she could persuade her friends to help her into the heavy gear and to stand up on the jetty to pump air for her.

✫

My friend Joanne recommended the diving instructors with whom she'd learned. I wanted a personal recommendation. Snorkellers and divers have died in the water, and in some cases negligence against dive companies has been proved. The dive industry has grown very quickly in Australia and safety procedures are not always observed, as has been shown by the sad case of the Lonergans, forgotten on the reef by the dive boat.

I did a PADI open-water course of instruction, PADI being the Professional Association of Diving Instructors, and "open water" meaning graduation from the pool to the real thing, open water, over the days of instruction. The course would qualify me to dive to 60 feet and to buy or rent scuba equipment. The training has three main components: classroom work in diving biology and physics, exercises in the controlled conditions of a deep swimming pool, and four scuba dives in open water. In my course the program content was controlled from the United States by means of video instruction and multiple-choice test papers. The franchised mechanized aspect of the PADI instruction was boring, though I guess we needed to know what *should* happen in the water before we took to the pool and did things the wrong way. I was also comforted that the video took into account that more older people, including women, are learning to dive.

One of my group said she had never imagined learning to dive would be so difficult—it looked so effortless in the movies. Australian Olympic athlete Emma George makes pole vaulting look easy, but that never took me in. I knew all along that diving would be difficult for me and it was. Simply putting the gear on was a struggle, at the end of which I'd be whacked, while others, waiting for me (as always they were a remarkably patient lot) had those precious few minutes to catch their breath. Falling backward into the water from the side of the pool was dead easy, though, so I was pleased to learn this is the usual way to get in.

I was definitely not a star pupil. Everyone else did everything more easily than me. I did everything wrong at least twice. At each stage of the learning process, I had eight pairs of eyes glued to my every stumbling action. If I'd started to gulp in water instead of air, I'd have been rescued at once, by too many people. In every class there is someone who is the absolute worst, and I filled that role. I made everyone else think they weren't so bad. But I persisted. I asked myself: What is the worst thing that can happen? Surely they wouldn't let me drown!

Teaching underwater involves special skills. Words no longer matter. My instructor Jason's gestures were brilliantly clear. The course requires the diver to remove and replace all their dive equipment while sitting at the bottom of the pool—a maneuver more useful in training for weeding out the faint-hearted than in an actual dive situation. Jason showed us at each step exactly what we should do. He'd punch his fist into his palm to get our attention. It made a sharp cracking sound. Then followed slow, deliberate movements in sign language, graceful and fluid. Air bubbled to the surface from the output of nine regulators, with Jason bubbling far more economically than the rest of us. How controlled his breathing was, like a singer or a yoga instructor.

We flooded the mask and cleared the mask. We took the mask off and put it on again. We all got bloodshot eyes, but we did it. I really thought making us share one air supply with two regulators was a bit melodramatic, but it turned out to be something I had to do in the open water, for real, a week later. We circled each other in soft, pool-blurred motion, the sun filtering down between us. There was one thing I got consistently right: I kept breathing throughout all the business of getting out of and back into dive gear underwater. Beginners tend to hold their breath in their efforts to maneuver their gear. Even as I released my extra-heavy weight belt and did it up again, I kept breathing. I knew my priorities.

One reason it was harder for me was that I continually felt myself at the limits of my physical fitness. I probably wasn't really, or I wouldn't have made it through to the end. Like childbirth, there is that stage of thinking "I can't go on," but then I'd summon up strength from somewhere. I had tried to prepare as best I could for diving instruction. I'd gone to the gym fairly regularly—well, twice a week. I swam a kilometer (about 0.6 mile) a day for weeks, with all that entails—waterlogged ears and perpetual bad-hair days. I had tried to be fit; my body just wasn't catching up with my will.

Where lithe young men required three weights on their belts, I had seven. Even so, I managed to keep floating more than I should have. When my buoyancy went wildly off underwater and I shot to the surface, Jason had to slip an extra weight or two into my vest pocket. One pool session I had two extra weights on one side, and only one on the other side, which tipped me desperately sideways. We were to take off and put on our fins at the bottom of the pool. The other skinny young things did not have to cope with being lopsided as well as everything else, I'd tell myself, snarling into my bubbling regulator and using up air at twice the rate I should have. I liked treading water, though. Give me a wetsuit and I just won't sink. Others had to work at staying afloat. I felt really good about the treading water part. I was the best.

But I'll not forget having a far-too-tight wetsuit, in the pool part of the course, something I couldn't do up myself, and my young buddy partner, John, having to help me. I decided early on that being embarrassed about myself was going to be totally out. Forget vanity; it would only get in the way of learning. It wasn't within my power to have John unembarrassed too.

We were a group of eight: four men, four women. The other seven students were mostly in their early twenties. They were nice to me. I wonder if I'd gone to a class for women of a similar age whether I'd have felt better. I doubt there would ever have been eight of us mad enough to enroll simultaneously.

I definitely was the buddy nobody in their right mind would want. Yet in the open-water section of the course, this paid off brilliantly for me and my new buddy Julie. Dean, the open-water instructor, more or less gave up on the task of including me in his group of eight. He persuaded Leisa to help out. Julie and I went off with Leisa as a group of three and had a great time together. One of the recommendations made after the inquiry into the deaths of Tom and Eileen Lonergan was that there should be one instructor to four learners. I certainly would have felt more comfortable with that ratio. With eight people, if more than one person is having trouble, the instructor has too hard a time of it.

I found it very hard to connect hoses and twist valves. My fingers felt weak. Things that should have clipped in easily on the buoyancy control device took me time. Learning through the body gets harder with age—it just does. There's a wide gap in there between understanding commands and carrying them out. Imagine being told to hold the weight belt in the right hand while rolling over onto one's back from right to left: with such a combination of orders the brain congratulates itself on getting the first part right. What was it to do next? Remembering to hold the belt in the right hand was important enough in itself. If the belt is held in the left hand, the weights slide off the end. I didn't drop the weights or the weight belt. (Later, out in the open water, I saw two instructors coming up from the depths with weight belts other students had dropped. I clung on tightly to my seven weights while others dropped their three. I gave myself a tick for that.) And—what next, after not dropping the belt? Well, blow me down; yes, it has to go back on again.

The fact is that body and brain tended to go in quite different directions. Three-hundred-odd years after Descartes described the split between body and mind, philosophers now say his mind-body split is an error: now it's all philosophy of the body. The school of body-is-all, all-is-body philosophers have never been diving. In underwater learning I felt it so clearly, this split into mind which understood the commands, and body which went its own way (upside down, and sideways, and in ways not possible in air). Then there was that something extra that hovered above both mind and body, observing both in chaos.

What would Jean Devanny have made of this? First, if she'd been shown the scuba gear, or in her case the Cousteau aqualung, she'd have leapt into it with alacrity. She wouldn't have bothered with diving lessons. Nobody did in the early 1950s. They put on the gear and went down, and some of them died. As I've mentioned, one famous pioneer in recreational diving on the Great Barrier Reef was the science fiction writer Arthur C. Clarke, who scripted the film *2001: A Space Odyssey*. Clarke had the same fascination with underwater space as he had with outer space. In the autumn and winter of 1955 Clarke went diving along the Great Barrier Reef. His book, *The Coast of Coral* (1956), tells the story. No doubt it helped that Clarke's diving companion, Mike Wilson, had been a paratrooper and frogman in England. I'm told that the worst person ever to be landed with as dive buddy is a bloke who tells you how much he loves skydiving, too. But a paratrooper and a frogman would be in a different category. Mike Wilson did both in wartime and survived.

Clarke and Wilson brought their Cousteau aqualungs to Australia. When they arrived, they found a new Australian technology they preferred—the "porpoise." The Cousteau aqualung gave a steady supply of air through a regulator, which allowed the diver to take down a tank of compressed air yet breathe normally. The Australian invention simplified the air regulator system and was more comfortable to use. It incorporated a buoyancy vest. Another device, a piece of string, allowed the diver to jettison the tank if he got into difficulty. An underwater parachute in reverse, this particular safety device is no longer used. If divers need to take off tanks in the water, they've practiced doing it often enough at the bottom of the pool. For adventurers such as Clarke and Wilson, diving school would have been altogether too soft.

Jean Devanny also had the oversupply of energy that divers need. "Eat, sleep, dive." That's what they do on dive boats. They get into the gear, go down, come up, get out of the gear, and do it all again, four times in a day. I was worn out, if exhilarated, after just one dive.

Diving is a highly unnatural activity, and it can cause death or illness and disability. We have to take our air with us, if we are to stay underwater for more than a few minutes, and we also have difficulty in dealing with the considerable changes in pressure that occur. As we descend, the volume of air in the lungs decreases and nitrogen, a normal component of air, starts acting differently in the body under pressure. While the body makes use of the oxygen in air, so that we exhale less oxygen than we inhale, nitrogen is an inert gas. This means that at sea level the amount of nitrogen inhaled and the amount exhaled are the same. As the diver descends, both the pressure and the amount of oxygen and nitrogen inhaled are greater than on the surface. The body uses most of the extra oxygen inhaled, but the extra nitrogen, under pressure, enters the blood and the tissues of the body, where it dissolves under the increased pressure. The problems known as "the bends" and "decompression sickness" arise from this excess nitrogen, which will dissolve out as harmful bubbles in the blood and tissues if a diver ascends too rapidly. The problem known as "nitrogen narcosis," or "raptures of the deep," arises, also from excess nitrogen, if the diver keeps descending, but it is a different thing.[5]

If nitrogen bubbles form in the blood and the tissues from a too-rapid ascent, they cause symptoms ranging from severe pain in the joints—"the bends"—to dizziness, headache, paralysis, and vomiting. The bends was the major cause of death in the pearl shelling industry at the beginning of the twentieth century. Now it can be treated by recompression in a hyperbaric chamber, where the nitrogen bubbles are reabsorbed by the body under pressure, and subsequent decompression can take place very slowly. Recreational divers are taught to time their dives so that they can "stage" their ascent, with rests at various levels, such as at 16 feet. With a slow staged ascent, excess nitrogen can be disposed of more gently by the body, without forming the dangerous bubbles.

Nitrogen narcosis is dangerous for different reasons. The diver who descends below the recreational diver's limit of 60 feet may find that at about 115 feet a state of euphoria kicks in. Nitrogen crosses the blood-brain barrier and penetrates the lipids or fats of the brain cells. It induces what is called the "martini effect," as it is like feeling suddenly drunk. Time ceases to matter. The normal survival mechanisms of worry about loss of control no longer work. The literature of diving is full of stories of divers at great depths who are observed to go deeper and deeper, out of control but not knowing it. The "raptures of the deep" have seized people who, in their rational moments on land, know about the dangers. Underwater they may immerse themselves in the flow of the activity, too much, too deeply, and forever. If the diver realizes what is happening and pulls out of it, the "martini effect" ceases dramatically at about the 120-foot level, unlike a real hangover from too much alcohol.

Canadian diving doctor David Sawatsky has researched the issues of diving and aging and says bluntly that older people have an increased risk of dying while diving.[6] There is an increased risk of developing decompression sickness with age, as well as a bone condition known as dysbaric osteonecrosis. An older person is more likely to have increased body fat, impaired circulation, damaged blood vessels, and damaged lung function as well as joint degeneration, and these factors increase the risk of getting "bent." Older people are also in greater danger from both hypo- and hyperthermia (becoming too cold or too hot). Sawatsky also says that an older diver should have the fitness level required to run a mile in eight minutes. Uh-oh to that one; I'm glad I read the article after the course was over.

Worldwide the number of people learning to dive is rapidly increasing, as the technology and the tuition methods become more user and client friendly. The PADI organization reports an exponential growth in certifications from its beginnings in 1967 to the 8.5 million divers certificated in the United States today. The number of older divers is rising, with the average age for divers now 36 years old. In the United States the gender balance is 72 percent male to 28 percent female. The rise in participation by women is a worldwide phenomenon, with the percentage of first-time PADI certifications to women rising from 26 percent in 1986 to 34 percent in 1996. In 1998 the PADI organization trained some 70,000 people in Australia.[7]

In the early 1990s the U.S.-based Women's Scuba Association campaigned for diving equipment to be designed especially for women. Before this time, manufacturers had been downsizing men's gear and coloring it pink.[8] Nowadays women can buy equipment that works better for them, and manufacturers find their women's lines sell well. Wetsuits now come in more than S, M, or L sizes, and snorkels and regulators that have smaller mouthpieces are available. The irreverent U.K.-based *Dive Girl* Internet magazine has a keen interest in underwater comfort and reports on topics that male-oriented dive magazines ignore, such as "dive bras."[9] With articles like "Seduce the Instructor: Good Move or Bad Move?," *Dive Girl* reads like a wetsuit issue of *Playgirl*.

☆

Dive magazines give a foretaste of the future of diving. Aimed at recreational divers, consumers of both dive gear and dive resort holidays, the magazines usually include an article in each issue on "technical diving," or activities presently beyond the capacity of the novice diver. Technical diving is diving with different air mixes, or with equipment such as the "rebreather," so that bottom time is extended and more dangerous activities attempted, such as deep dives or cave dives.

The technical diving article in the dive magazine is the article that tempts the reader on to imagine a new personal frontier, technically aided. The prediction is that in 15 years' time, gear that is now for the specialist will become routine. Nitrox is oxygen-enriched compressed air, or air to which oxygen has been added, so that it goes from being 21 percent of ordinary air to 32 or 36 percent of the oxygen-nitrogen mix. Nitrox has less nitrogen, which in turn means that the body will absorb less nitrogen. Both the time spent on the bottom and the number of dives a day can be safely increased. Nitrox lessens the risk of decompression illness. Rebreathers use technology in which the same air is rebreathed; the diver exhales into the equipment, which scrubs the air free of carbon dioxide and recirculates it. Rebreathers are used in underwater professional photography, as there are no air bubbles. Rebreather technology is still very new and complicated, but it promises to revolutionize diving, making gear much lighter.

Soon there will be navigational instruments such as position locators as part of the dive computer. Diving will be more accessible for older people. U.S. industry leader Bob Clark predicts that "the diver of the future may be from the 'gray group'—people in their sixties who are in good shape and interested in picking up diving for the first time." A gloomy prediction comes from diver Howard Hall, who says that "the tech diver's need will greatly influence regulators, harnesses, mixtures of breathing gases, and rebreathers. The disadvantage is that a whole lot of tech divers are going to get killed. But that's their right as far as I'm concerned. If they want to do a 500-foot dive in 30-degree water, that's up to them."[10] Lest our 60-year-old diver be led astray to the 500-foot dive, a better idea might be to own your own personal submersible, an underwater equivalent of the snowmobile. The Bellaqua BOB, or Breathing Observation Bubble, may be ordered for a mere $9,995. The ad promises: "Dispense with cumbersome scuba gear to cruise in silence amongst reefs, wrecks, and fishes. Spectacular visibility is achieved using a clear, acrylic, air-filled dome mounted on the head piece."[11] BOB is a like a motor scooter with a space helmet attached. In the future a trip underwater may be roughly equivalent to a Saturday afternoon drive in the country. One day will there be underwater traffic lights?

I won't ever go cave diving or descend into the abyss, but I'm not going to worry about that. My experience, awful as some of it was, was still brilliant, if brilliantly awful. My first ocean dive brought to mind movies in which the hero is swallowed by an intergalactic monster and finds himself tumbling through an alien digestive tract. Life in all its lumplike or tentacled or fanlike or billowing forms passes by. Small schooling fishes swim above and around. They come this close and no closer to the diver and turn at a certain distance as if encountering an invisible force field. Fishes feel the turbulence we produce, as our skin feels the pressure of the wind. Flocks of fishes glide in an underwater sky. Down the bottom, skulking around coral in the sandy lagoon floor, are tiny damselfishes. They compose themselves in layers, here a blue group, there an orange group, finning slowly though the branches of uncolor-coordinated pink coral. The next layer up are the yellow-tailed fusiliers and the next the yellow-banded fusiliers. Underneath the dive boat lurk sweetlip and trevally. Large batfishes look at divers with much the same regard as divers look at them. To see all of this underwater while keeping an eye on the air-pressure gauge and maintaining buoyancy at a wobble is a real achievement.

I tried to look for trochus shells. I had given myself some preliminary practice at this in the Great Barrier Reef Aquarium where I could always find them, no problem, in the smaller aquarium tanks. Out on the reef I didn't have a hope. The trochus shellers used to feel for them with their fingers. That will no longer do. Poking fingers under coral ledges is not a good idea, either for fingers or for corals.

You don't have to have the time, the tenacity—the sheer madness in my case—to do a whole scuba-diving course, but I think visitors to the reef should make at least one dive. Most people who do not have specific health problems may take a one-off trip in the "resort dive," the one-to-one experience with a diving instructor who checks up on the technical details and smoothes the way down. After I had done my training, a friend I met on a reef trip, Marie, told me that I really should have had one-on-one instruction for my dive course and told me where I could get it, on Magnetic Island near Townsville. She reckons the "personal trainer" approach to diving is more appropriate for oldies. She's right, but I wouldn't have had as good a story to tell.

## POSTSCRIPT

Three months after I wrote the words above, I went diving again on the Great Barrier Reef. Going out as a qualified diver—if just barely—proved much easier than doing the PADI certification. I solved the problem of buoyancy control easily by wearing a Lycra suit rather than a full wetsuit, which for warm tropical waters in the middle of summer was perfectly adequate, at least for me. All I then needed was four weights. The air tanks I used this time were larger, so that I always had about 100 bar (a measure of air pressure) left over at the end of the dive, compared to the bare minimum of 50 bar while on the course. Because the air-pressure gauge is not exact, the advice always given is to never let it go so far as to register empty and in practice never to let it get below a reading of 50 bar. After I was qualified, and with a larger tank, some of the anxiety about running out of air left me. I could maintain buoyancy more easily, with the extra practice. I had more time to look around and enjoy what I saw. I could forget all the diving technology I carried and immerse myself totally in the flow of the ocean, the flow of the moment, the flow of underwater life.

# Fishness

The day heats up. The surface of the water lies cool, calm, and inviting. I slide in. Fishes soon find me: two diamond-shaped trevally of the reef shallows. If I am lucky, I'll see the gobi and the shrimp keeping house together underneath a rock in this sandy prelude to the reef. The gobi is a tiny sandy-colored fish that shares its burrow with the alpheid shrimp. The shrimp does most of the housework, digging the sand out with its claws and generally keeping the entrance hole clear. The snorkeller has an excellent view from a yard above. The small shrimp works hard in the face of waters always ready to swirl the sand around and back again. Keeping watch, the fish guards the shrimp from harm. When the shrimp sees the gobi turn tail and dive back into the burrow, it too dodges back to safety. The fish darts in and out, dodging the busy shrimp. It is a thoroughly companionable relationship.

The first underwater experience gives both less and more than television documentaries promise. Colorful corals, hugely magnified, big fishes, sharks, even the hated crown-of-thorns starfish already exist in memory and so in anticipation. Real corals seem muted in color, and smaller, because the polyps are retracted and mostly invisible in daytime. The surprise is the fishes. Hundreds of tiny damselfishes dart around in shoals, finning their swift way through sunlit waters. Azure damsel fishes, only an inch or so in length, capture all possible shades of blue in one sleek swift body. They sweep around the intruder and onward, keeping to the path the school is collectively traveling.

The first experience is both strange and familiar from underwater film. The best simulation of being under the sea has come, for me, with the experience of the 1999 Imax 3-D movie *Into the Deep*. The viewer relaxes into the theater armchair, wearing a visor headset that helps create the 3-D image. Fronds of Californian kelp forest materialize in the cinema and float toward the viewer and beyond. A shark cruises out of the blue into the cinema and noses up to the headset wearer. Imax 3-D conveys the underwater world with a heightened reality. If I went into the Californian kelp forest, I know I'd not see a fraction of this beauty. I'd be just one person, not a team of expert photographers selecting their best shots.

Yet still the cinema simulation misses something, even though now with 3-D technology it may attain near-perfection. In the reef experience, in actually "being there," the multiple aspects of yearning collide with sheer physicality. The heady mind-body mix stimulates the sense of wonder to overload. Sensory information is fed to the swimmer from all directions—up, down, sideways—differently from on land. Smell and taste become irrelevant, or at the very least, the taste of plastic in the mouth is mercifully diminished. Though divers hear the disconcerting sound of their own breathing, magnified, the sense of hearing is dulled for people but not for fishes. We hear through our ears; they sense sound three ways, through their air-filled swim bladders, their unique "lateral line" sensory systems, and otoliths (earstones). Weird, too, is the sense of touch underwater. Warm water flows over exposed skin and creates a liquid cocoon, a personal comfort zone that flows along with the swimmer. Otherwise it's look, don't touch: so many jagged rocks to cut the skin, so many poisonous spines out there to jab.

It is the visual that overwhelms. The fact that sight is restricted to the field of view of the mask adds to the effect. What is that just outside the field of vision? Is it bigger than me? As I observe this corner of this world, what is observing me? Beyond each rock and crevice new marvels are revealed; possible terrors are concealed. Other lives are lived in this other dimension of existence in a manner totally alien to our own. Normally we are their predators—we eat them; but we are the strangers here. Some of them are bigger than us. The teeth of sharks are magnificently adapted to catching, holding, and biting prey. The sense of wonder is also a sense of wonder at being alive in a place where so many dangers lurk.

☆

"Fishness" is a word coined by Jeremy Tager of the North Queensland Conservation Council, as he makes the empathic leap to wonder what it is like to be a fish. He is forced to conclude that "understanding the nature of fishness may exceed the limitations of the human imagination, creativity, and empathy."[1] The quantum leap is to imagine a consciousness with different states of awareness. How might fishes experience tides and currents, night and day, territory, food, reproduction, and death? How might they experience the reef, and how might their experience be different if reef corals grow from shipwrecks or from piles of old tires with which artificial reefs are now being created?

One path to imaginative understanding lies in the walk through the Great Barrier Reef Aquarium in Townsville. It was there that I met my one true invertebrate love and we got some empathy going—at least I got it going. I don't think it was reciprocated, but, hey, you've got to start somewhere. It was a red-lipped stromb of the family *Strombidae*, a small marine mollusk that carries a bright orange and white shell on its back. It sat with its plump foot pressed firmly against the glass side of the tank at the entrance of the aquarium. It was totally blissed out, this stromb. Eye-stalks, foot, and siphon were stretched out of its shell to the limit, all systems go, eyes swiveling, siphon vacuuming, foot lurching from side to side. What caused this moment of joy was a strip of green algae in a hard-to-get-at corner of the tank. Tucked in behind a fat aeration tube was some green growth its human minders couldn't reach to clean. This was a stromb in which the hunter's spirit had not been dimmed by the soft life.

Ours was a relationship made possible by modern aquarium technology. If I'd gone out on the reef with a portable glass tray, the kind reef photographers use, and with the patience to sit and watch, I might have seen some of this. But I'd have missed the stromb's passionate mission for algae. Out on the reef I'd have been unhappy about picking up a stromb because of its close resemblance to the lethal cone shell. The cone shell has modified teeth, like small poison-loaded harpoons, which it shoots out if disturbed. It kills its prey, including humans, fast. At the aquarium there's a cautionary motto they teach the children: "If it's a cone, leave it alone." They have harmless strombs in the children's touch tank with an instructor hovering over to say: "Don't do this when you're alone."

The two eyes of the stromb stand at the end of long stalks and swivel in contrary motion. While one tiny blue eye is on the watch for the next best slurp, the other eye checks up on its back, or its sides, or underneath, a good idea if you live in a place where predators may strike from any direction. This stromb was not to know it lives in a predator-free tank. Staff don't want the exhibits to eat each other. There is even a notch in the lip of the stromb's shell so that the right eye gets as much room to move as the left. Watching the stromb, I couldn't tell what was more important to it, the sight or smell of the algae, but however it knew, it knew what it wanted and was single minded about getting it.

The aquarium is inspired by the human desire to linger in curiosity, and not only enjoy the beauty of the creatures but also to sustain the pleasure they evoke. An otherwise fleeting moment of underwater delight is thus artificially maintained and experienced anew daily. What may usually be seen only with difficulty or not at all becomes widely accessible. I am privileged to get to see this stromb, here on this artificial reef to which live corals are attached, where the fishes are maintained in good health by technologies that use algae for water purification and by the artificial separation of predator and prey.

I was delighted by its delight in food, its opportunism, its mastery of every appendage of its body, all the humanlike attributes I could relate to. Empathy must start somewhere. There is also the question of getting to know what you are by what you are not. Humans do not wave their snouts around sucking algae, but I can see some of the advantages.

In the aquarium, guides tell fond stories about the fishes, the sharks, and the turtle, and they give them names. The green turtle is Lucky; the leopard shark is Leo; the potato cod is Humphrey. The huge Queensland groper is Dom, after the aquarist who trapped it. Dom is also short for Dominator, as this fish definitely dominates the predator tank. Dom has its territory near the surface, where it languidly floats with the occasional waft of its tail while keeping a sleepy eye on things. At feeding time the guides tell stories about the predators' personalities. Lucky, the green turtle, is very curious. He rushes to get a fish at feeding time, even though green turtles normally eat sea grass. "No one ever told him," someone comments. Lucky is lucky because he's the one in a hundred turtles that survives to adulthood. Recently the predator tank has been reefscaped so that it simulates the *Yongala* shipwreck. The aquarium staff recount how curious the predators were about the changes the divers were introducing into their tank. Lucky kept putting his nose into the replica of the ship's safe, left on the tank floor as it was found at sea.

People commonly think animals are like themselves and describe what animals do in stories that support these interpretations. This is anthropomorphism and is considered a trap for students of animal behavior, to assume that animals have the same kinds of feelings as people. Yet somehow it happens that we do attribute to animals motives similar to our own, and indeed we are perfectly prepared to extend these notions to fishes. Telling stories about fishes is an activity long practiced by fishermen, who will attribute intentions such as cunning or curiosity to their fishes. The shark is associated with evil, man-eating intentions. Now, possibly serving the interests of reef tourism, visitors are encouraged to regard at least some sharks, the black-and-white-tipped reef sharks, as creatures that just want to be left alone—they have this virtue, where once they had that vice. Anthropomorphism, in telling stories about animals, is neither right nor wrong in itself; it is just something people do. It is one path along with others to a partial understanding. Watchers of whales, dolphins, and dugongs also give names to these large mammals, while assuming they do not name us in return. (The guides hadn't named the stromb. What seems appropriate for large animals would seem silly for a mollusk.)

The aquarium is a reminder of the complexity of modern life, especially the technology it takes to maintain nature artificially. The aquarium visitor also marvels at the machinery that sustains the spectacle. In the Great Barrier Reef Aquarium, an "algal scrubber," a machine for maintaining water quality, is on show beside the children's touch pool. Water from the aquarium tanks is pumped over the top of rafts of brown algae, which feed on the nutrients in the water and help clean it naturally, as they would out on the reef. The stromb in the high-tech tank is like the human in the high-tech city—the survival of both depends more and more on technological expertise. The aquarium tank recycles air, water, and some food, just as a sustainable development incorporating recycling is called for on a larger human scale. A machine that uses algae to clean water harnesses natural processes toward an end that benefits algae, fishes, and humans.

I like to walk around the aquarium and marvel at how clever engineers and marine biologists are to create and maintain a corner of perfection and to make it so accessible. Relationships between humans, nature, and technology are changing with great rapidity. New things become possible, calling old assumptions—such as the reef always being there, for exploitation—into question. The aquarium makes it possible for the viewer to appreciate fish behavior and perhaps grow fond of it, as fondness has grown for the exploits and "character" of the domestic dog and cat. Add in the technology of the movie camera and time-lapse photography and the hitherto private lives of fishes are exposed and magnified for us in all their drama.

When the first large marine aquarium was built in nineteenth-century Britain in Brighton, the science journal *Nature* published gossip about the new aquarium's residents. The two porpoises were settling in nicely, but, alas, the new octopus ventured too far out of its nest of oyster shells and got eaten by the spotted dogfish.[2] The policy of separating predator from prey must have had to be worked out by trial and error. Apart from the inherent tragedy of this aspect of the fishy condition, it gets expensive when the inhabitants eat each other. Today, *Nature* is the most prestigious and austere of all learned scientific journals. Gossip about fishes never rates a mention.

The stromb and I were at the still small center of a web of significance spun between nature and technology, nature and culture, humans and the rest of the organic world. Of the two of us, I was the only one in the position to know it. Our relationship spun out into all the connections, past, present, and future, between people and the place from which the stromb had come, the Great Barrier Reef of Australia. The aquarium is its home as the result of decisions made about conservation of reef resources, tourism, and public education. The reef itself was its home as a result of shifting continents, migrating corals, the coming and the goings of the last Ice Age. Eons of geological and biological activity led to this one chance encounter, of one human with one blissed-out stromb caught in the act of siphoning algae.

With the invention of scuba gear, marine biologists now hover above their subjects and observe the rhythms of their lives. The fish has the advantage on the diver. The diver must carry air and control buoyancy artificially. The diver has a restricted field of vision, hears few sounds other than bubbles and breathing, and tastes and smells nothing. The diver feels a sense of wonder and assumes the fish does not. The fish, on the other hand, balances easily in the water by means of an internal float of gas. Its body is sleek and streamlined, with muscles especially adapted to give a good forward push in water. Fish eyes project and allow a wide field of peripheral vision, with depth perception possible above, below, and in front. The diver sees only in front and must turn to see what lurks elsewhere.

As fishes school, they move together as if choreographed with a dancer's sense of unity and grace. Their lateral-line sensory system reads signals way beyond the limitations of the human senses, detecting currents and changes in currents and pressure. For fishes there is no such thing as the silence of the deep. At night the reef buzzes with the noise of shrimps.[3] As the diver observes the fish, the fish senses the diver. The fish has the advantage—as long as the diver comes in peace, unarmed.

To gain some grasp on the nature of fishness, the diver must try to think through what it must feel like to be an individual fish in some way always aware of the presence of the school. The philosopher Thomas Nagel asked a related question of a different species: What is it like to be a bat?[4] After thinking the question through, he decided that he could never know what a *bat* felt like being a bat. He could only arrive at a conception of what it felt like being a human with an extra sense of echo location, of knowing where he was by sensing messages beamed back from surrounding obstacles. Both echolocation in bats and the lateral-line sensory system in fishes are ways of knowing the world that are alien to humans. Thomas Nagel offers no hope to Jeremy Tager in his quest to understand the nature of "fishness." Nagel says, in the case of bat sonar, that this sense is just so different from any that humans possess that we shall never have insight into the inner life of the bat. Even though navigation devices for the blind use a device similar to echo location, this is not the same thing as having the sense embedded in one's neurophysiological constitution.[5] The day when human divers can move like fishes in underwater schools will never come.

We know that fishes have a form of 3-D sideways perception. We can probably also say of fishes, as Nagel says of bats, that we believe fishes feel some versions of pain, fear, hunger, and lust, where lust seems an appropriate term for the remarkably intense and unceasing effort devoted to courtship, acquisition, and defense of mates in most species of coral reef fishes. What seems the impossible next step, though, is to experience subjectively what it is like to be the fish or the bat. Yet we can say that fishes and bats have their own species-specific viewpoints of their world, and it could be that their world is experienced just as richly as our own. We think, though, that we humans have an edge on them, because we, uniquely, can have concepts of what the world is like. Fishes can't imagine the earth from space.

There seems no way of erasing away the human point of view and substituting that of the fish. Yet it is worth a try, to imagine what it is like to live in a state of different sensory awareness, without the capacity for self-awareness but with a kind of school awareness. Perhaps it would be like having a sense of community without a sense of self. Anthropologists who study human communities sometimes use the word "dividual"—"individual" with the prefix "in" removed—to indicate social groups in which the sense of community is particularly strong. Fishes, as they move in schools, move as "dividuals" all. They show there are ways of living cooperatively in groups that humans might like to ponder. There was a time in the past history of humans on earth when they first learned to gather in groups around a campfire. In the future, people will gather differently together. We're not going to end up fishes, but we may end up relating differently, and possibly better, as dividuals all.

Sometimes the impression comes most strongly, whether underwater or in the aquarium, that as I am observing this fish with interest, so this fish, in return, is observing me. In this encounter, eyeball to eyeball, who knows what the fish is thinking, as indeed, if I similarly encounter another person, I cannot read his or her mind. Peter Chermayeff, a designer of cutting-edge aquariums in the United States, goes so far as to say: "There's something utterly primal about the direct contact. To look a fish in the eye is a spiritual experience."[6] The aquarium designer thinks of the fish as having value beyond its practical relevance to his work, as intrinsically valuable in itself. The return of the gaze means something. It acknowledges that a fish-human encounter can have a two-sided meaning. Fishes are a valuable source of protein for many of the world's peoples, but fishes mean something more to the world than an element in a design or a good fish dinner. "Fishness" is, in itself, something many people value. The conservationist argues that "fish-kind," in all its multiplicity of species, needs to be preserved in the world, for more than its usefulness to people as food.

My next question is: But can I bring myself to extend the benevolence that I feel toward the stromb and the fish to something a lot less nice—the crown-of-thorns starfish? The crown-of-thorns starfish has a bad reputation as a destroyer of coral reefs. It is large—dinner-plate size—and covered with mucky green-red spikes. It has the habit of extruding its stomach over living coral, sucking the polyps into its capacious gut, and moving on, leaving a bleached and lifeless skeleton behind. Outbreaks of crown-of-thorns serious enough to be called plagues have blighted the Great Barrier Reef in recent years, leaving some reefs devastated, mere shadows of their former glory. Can I feel fellowship with the crown-of-thorns, or with any creature that, in minding its own business, threatens mine?

Personally, I wouldn't eat the crown-of-thorns starfish, so we can't even have a utilitarian food relationship. In fact, the starfish has few natural predators, except for a few large shellfishes, and that is part of its success in survival. I do not find it beautiful, though it can be made to look so by clever photography. Images of the juvenile photographed from underneath as it sticks to the bottom of a clear glass tray show a lacelike radial symmetry, which delights—until I read the caption underneath. It is very capricious, this love-hate relationship with other species.

I have to take a further step, to understand the nature of fishness not only as an individual attribute but also in relation to reef communities. Polyps exist in colonies. Reefs exist as a congregation of coexisting species, many of which eat each other. There is fishness and there is crown-of-thorn-ness and there is reef-ness. The individual fits into a web of reef life, a life that is always in a state of change. Now one set of organisms gain ascendancy, as with the coral polyps that build the reef, then others take over, and it may be those organisms that destroy. The flip-flop mechanism of dominance is neither good nor bad in itself but simply the way things are.

One way to approach the next to impossible task of imagining reefness is to begin with creatures that have a special status in human eyes, large marine mammals with which it is easy to empathize. Dwarf minke whales have not been hunted for food, being small and swift, and anyway only recently described as a species. They grow up to 25 feet long. In the winter months of the early 1990s dwarf minke whales began approaching tourist boats on the Great Barrier Reef. The whales seemed to be curious about people. Soon a north Queensland tourist venture, *Undersea Explorer*, brought people out to Ribbon Reefs to swim among the whales. The whales approach the people; that is the significant element in the encounter, and people feel delighted that they, in particular, are chosen.

What whales make of the encounter is unknown. It is humans who say the whales are curious. People value the whale experience. Whales turn up to see people. Can we say in turn that whales *value* their human experience? They can't show us they are going though a process called evaluation, but they do seem to be showing by their behavior what they "value" or are interested in. They don't like people diving down or swimming toward them. They like humans to behave in a controlled and more predictable way. That is why zoologist Alistair Birtles recommends that people string themselves out along a rope on the surface.[7] This way the whales control the approach to the swimmers. It is probably not stretching it to say that what the dwarf minke whales are exhibiting is something like our own experience. Something odd is happening—interesting, yet not threatening, and they want to know more about what's going on.

A similar experience of animal curiosity is reported by dugong researchers in Shark Bay in Western Australia. Normally, dugongs are shy and retreat from contact with humans, a wise move as they are often hunted for food. Yet the zoologist Paul Anderson and his colleagues reported an unusual experience as they snorkeled in the midst of a large herd of dugongs. At first the swimmers were approaching the herd from a boat at anchor. As they swam toward the dugongs, they were delighted to find that they were being met halfway. The normally shy animals were swimming toward them. The dugongs would approach, singly or mother and calf together, to circle beneath the snorkellers. The animals allowed the swimmers to dive quite close, to within a few yards. One dugong circled and was gone, to be replaced by another animal. As the scientists observed the dugongs, so they were observed in turn. After a while scientists learned to identify some individuals by their scars or their barnacle attachments.[8] Whether dugongs identified some of the scientists is unknown.

In thinking about marine life it is possible to imagine a continuum running from people to dugongs to sea turtles to damselfishes to coral polyps to sea grasses, in which there is a transition from fully conscious human experience to the totally unconscious sea grass experience. Coral polyps do more than feed and reproduce. They have in place adaptations and survival mechanisms that are of value to them and to species that have evolved alongside. The coral even has a primitive system of defense of territory. When the polyps at the margins of a coral discover a competitor, the coral acts to defend itself. It projects its digestive strands, the "mesenterial filaments," through the wall of its body into the outside world. These strands wave about in the contested space, secreting their digestive juices just as they did in the gut, and when they find the intruding polyps, they eat them—if they are from other species. They do not eat their own. When they feel under pressure from competitors, the polyps at the margins grow specialized "sweeper" tentacles with large stinging cells. These emerge at night and clean up the surrounding territory. The coral polyp defends its life and environment without knowing what it does. It is part of a continuum along which humanity lies. There is a bit of the coral polyp in us all.

There is a richness in the system of the reef, in its conflict, its resolution, its states of equilibrium, and its states of chaos. The reef system recycles its nutrients and its energy. There are few nutrients on a new coral cay, but birds get busy and soon provide phosphorus and nitrogen through their guano. Zooxanthellae enter into a symbiotic relationship with corals and add their energy to its sum total. The complexity and interdependence, richness and diversity, of the reef have meaning for human lives that is intellectual, spiritual, and aesthetic. The reef also has economic, recreational, historical, and scientific value, and more recently we are coming to appreciate that it also has value by virtue of its genetic diversity.[9]

This richness is available in the large tropical reef aquarium in a way that it is not with a zoo. In the zoo there are no systemic connections. The animals are fed at publicized feeding times, and we take pleasure in observing the relationship between the keeper and the kept but not between the kept and the natural system. In the aquarium, apart from the artificial separation of predator from prey, it is still possible to see interspecies relationships in action. Herbivorous fishes find their own food in algae growing on fiberglass walls. Damselfishes guard their territory of branching *Acropora* corals, as they do in the wild, and anemone fishes maintain their symbiotic relationship with their anemone host, which guards them from predators in return for luring food fishes into its tentacles. In addition, the fishes have learned an extra thing or two about their unusual environment. They know there's a talk show on at 10 in the morning and that a diver will appear with a squeeze bottle of marine fish flakes. They cluster around the plastic bottle in a fish cloud of color that brings deep pleasure to the viewer on the other side of the tank wall. Viewers don't see the artifice; they are immersed in the living relationships.

# Going with the Flow

Sydney's Darling Harbour aquarium at night has its own peculiar fascination. In this most artificial of natural environments, fishes still know the rhythms of their bodies and come seven o'clock they are looking for a place to sleep. Their large tank is not the bottom of Sydney Harbour, just a replica of it, and the sea creatures make do as best they can. Green turtles seek crevices to hide their heads. They nose away at the thick carpet cover that lines the sides of the aquarium until they're well tucked up, even if the carpet is coming loose. Sharks drift dreamily overhead. They don't need to hide. They're always safe. Rays settle on the sandy slopes closest to the air bubbles. A school of diamond-shaped gray fishes, yellow-streaked *Monodactylus argenteus*, lines up close by a large wooden pseudo-jetty pile, noses to the wood, bodies each angled in the same direction, to . . . what? The pile or the flow of water through the tank? Whatever it is, they like it that way, and bleary-eyed and torpid they fin, in just enough slow motion to keep themselves in accurate alignment.

Going with the flow. Effortless action. Flowing in harmony with natural rhythms. Out on the reef part of the attraction is slipping into warm tropical waters and letting oneself go, yielding to and not resisting the forces of nature. A drift dive along a steep drop-off sweeps the diver beside a wall of coral to one side, the deep blue of nothingness to the other. Or is it? From out of the blue a dark shape slides into view, then just as easily, just as swiftly, slides back into the gloom. Other forms of life, some of them larger than me, and with bigger teeth, are also drifting along this wall. They do so with minimal effort, with an occasional wriggle of a flank and a sideways motion of a tail. The diver maintains buoyancy and lets it all happen. Drifting is the life. Great beauty lies all around. Let it flow past.

Drifting with the flow is an everyday fact of life in the ocean. The ambient medium is always in motion, and sea creatures harness its power and direction to their own animal ends of survival. It is different for most land-based creatures, though spiders may drift in air as well as microfauna, and sea eagles, frigate birds, and glider pilots. For the most part, our feet are on the ground, and wind plays little part in everyday life.

Modern reef-building corals evolved some 200 million years ago with the symbiotic relationship between corals and algae. In the warm waters of the Tethys Ocean, the ancient seaway that opened up as today's continents drifted apart from each other, the tiny unicellular algae first penetrated the tissue of the coral polyps and grew accustomed to life inside their hosts. The algae contain chlorophyll, which converts sunlight, carbon dioxide, and other nutrients into oxygen and energy in the form of carbohydrates. A symbiotic union is one that is of benefit to both host and guest. Here the algae, also called zooxanthellae or symbiotic dinoflagellates, gain protection from predators and find nutrients. The coral polyps gain oxygen and energy from the algae, and this extra efficiency helps their skeletons absorb calcium more easily.

Over tens of millions of years the Tethys Ocean closed up, as present-day India collided with Asia to form the Himalayan Mountains. The only surface remnant of this ancient seaway is the Mediterranean Sea, though its relics are found in the rocks beneath the earth. Coral species were widespread from the current Americas to the Mediterranean to Southeast Asia. The continents drifted apart, and the corals colonized new places, harnessing the huge forces of the ocean to their own ends of survival. Coral polyps lead a sedentary lifestyle, encased in their chalky exoskeletons. Yet theirs is truly a "rags to riches" story, in which a plucky little life form makes good and achieves global expansion ambitions, to the point of dominion over all the shallow tropical seas of the world. The polyp is so tiny, yet *en masse* it produces huge limestone structures. Coral has been measured at 4,300 feet deep at Enewetak Atoll, and the Great Barrier Reef is 1,550 miles long.

One day on the reef I was enjoying the glide of the ship through the water, the warmth of the wind, the freedom of the seas. We were far away from dirt and dust and cities and pollution. Something broke the spell: it was scum, frothing away on the ocean surface, long lines of it, rather like the stuff regurgitated by a blocked kitchen drain. But this scum was not a result of human activity, not this time. This was organic, living scum, the spawn produced by corals. This scum was the seething evidence of invertebrate sexuality—produced when corals and mollusks and worms ripen sexually and discharge their gametes. These minute organisms are discharged from all over a reef: How come they collect together in these long lines of froth? This scum is getting itself together, to go with the flow.

Mass spawning on the reef is one of the marvels of nature. On one night of the year, a night determined by cycles of the moon and the sun, some 130 species of corals spawn on the Great Barrier Reef. No one yet knows how they tell time so accurately. The fact of mass spawning was first recognized in 1984 by a team of scientists from James Cook University. Since then naturalist-photographers have been busy underwater capturing these moments of synchronized spawning for reef documentaries and what was unknown a few years ago has become a familiar image on our screens.

David Hannan's video, *Coral Sea Dreaming*, shows polyps magnificent in their night glory.[1] Out of their open mouths bright-colored, fringing tentacles curl toward the midnight blue of the ocean. Normally these delicate fringes would be busy sweeping in plankton for food, but this one night of the year they are preoccupied with a more primal urge. The polyp mouth expands and, like childbirth in miniature, a tiny mucous-enclosed package is squeezed into the water. Afterward the tentacles resume their gentle wafting motion. Separate egg and sperm packets rise to the surface of the sea and, as they burst open, the gametes are thrown together in a reproductive soup. So many corals, so many gametes of different species. Yet like meets like on the surface of the sea, right gamete connects with its appropriate other, and new mobile coral life forms, the planktonic *planulae*, are created. They will travel around for up to a month until they find a space to settle and grow, if they are lucky, or if they do not survive, they die as food in other hungry mouths.

*Coral Sea Dreaming* takes the viewer on an exhilarating ride with the coral gametes, as they shine like galaxies of stars in an underwater night sky. The small gold spheres rise from the mouths of the polyps through the dark blue sea up to the surface. Then the transition is made to the view from above the water. Scum! The glorious galaxy has dissolved into unlovely dishwater froth. It is now day, and the coral *planulae* appear as a vast spread of scum over the surface of the water, swirls of brown and white against the clear clean blue of the Coral Sea.

Next cut. The view is from high in the sky. What seemed like scum now becomes clearly visible as the pattern of reefs in the sea. Clever David Hannan. With the skills of the film editor's craft, he has turned beauty to ugliness, ugliness to beauty again, and has fully confounded my aesthetic prejudices. Sometimes scum is beautiful. From incongruous scum grows the great structure of the reef. The reef is a place of natural wonders. This scum is where it all begins.

When a coral larva settles, it excretes a bony skeleton and cements itself to a base. The other end becomes a mouth complete with tentacles and stinging cells. Soon it will clone itself through asexual reproduction, budding out other polyps that in turn form their new skeletons around them. The little coral polyps bud away and leave their skeletons behind. The two forms of reproduction, sexual reproduction and asexual reproduction or budding, make for great efficiency and flexibility and are characteristic of the life histories of many other reef organisms. Asexual reproduction means corals are not totally dependent on the success of their "one night of the year" sexual reproduction strategy. Colonial animals have a "reproductive plasticity" that humans can only wonder at. What must it be like to be both an individual and a member of a colony of identical individuals? Reef scientists Robert Buddemeier and Robert Kinzie have wicked fun with this idea. They say that human researchers may have an unconscious bias toward valuing sexual over asexual methods of reproduction and are "lookin' for love in all the wrong places." Meanwhile, corals bud on, and both their sexual and asexual activities provide evidence of reproductive success and hybrid forms that continue to puzzle geneticists.[2]

Since the mid-1980s surprising news has surfaced about the wide dispersal of fish eggs and larvae from their sources in one reef place to where they may eventually end up, several reefs downstream. Most fishes shed their eggs and sperm directly into the water, where fertilization occurs externally (as with corals). Reef fishes are most prolific with their eggs, releasing large numbers every day over their period of spawning. The fertilized eggs are soon swept away by tides and currents into the open ocean outside the reef, where there are fewer predators. The eggs swiftly develop within a few weeks from fish larvae into tiny juvenile fishes about a centimeter (about two-fifths of an inch) long. The larvae and juvenile fishes may then travel tens to hundreds of miles along the reef system from their place of origin to a reef where they will "settle," that is, grow to maturity. Researchers now are able to establish the distance and direction of movement of the larvae and juvenile fishes and are making some surprising discoveries. For a start, they are quite zippy little swimmers and seem to be able, at a very young stage, to sense where a reef is from about two-thirds of a mile away. If they are not swept downstream to another reef, they seem to be able to return to their reef of origin when they have grown a bit more.[3] Survival depends on many interconnected factors. In "good years," reports ecologist Terry Done, about 10 to 100 times more juvenile fishes reach coral reefs than in "bad years."[4] Populations fluctuate, but a decrease in one place may be related to an increase on a neighboring reef. The extent to which juvenile fishes seem to know where they are when they go with the flow is important for establishing sustainable reef fisheries. The trend is toward establishing reef "refuges" for fishes, in reefs that are good "sources" of fish larvae, and to make these places off-limits to all fishing.

From the Tethys Ocean the corals spread out in the Indo-Pacific region until they found a place they liked, and there they stayed. Corals drift in a variety of ingenious ways. Their existence as larvae is short term—from two to four weeks—so they can't take off over vast stretches of open ocean, at least not unassisted. Some go rafting on pumice stone from volcanoes under the sea. A slick of coral larvae encounters pumice (or bottles or other flotsam), and the larvae settle on it as a new home. They extrude their limestone bases and cement themselves on, and soon clone themselves into new coral colonies, which can raft from Tonga to Fiji and New Caledonia. After about a year at sea they may reach the Great Barrier Reef. The coral accretion gets larger and heavier and gradually sinks deeper into the water. If it sinks too deeply in the ocean and too far from the light, it will die. If the stone rafts onto a reef and the corals survive, they disperse themselves widely through the oceans and contribute to coral diversity.[5]

Drifting in the ocean turns out to be a surprisingly organized affair. Larvae and flotsam and jetsam, pumice stone and oil spills, all collect in patches on the surface of the sea. The surface of the ocean may be still yet underneath may be waves that are hidden, waves that do not quite peak at the surface. The hidden waves push floating stuff together into long lines, whether it is driftwood or pumice stone, coral larvae or oil spill. *Windrows*, they are sometimes called, in comparison to the lines of chaff left in a hay field after harvest. Something is happening, though it looks as if nothing is happening. There is flow within flow. The small and weak, the coral larvae, are harnessing the power of the strong, the ocean forces—not, of course, with any conscious intention of doing so. Rather, it has happened, in the course of the evolution of life in the oceans, that the power of dispersal of ocean waters has worked to the advantage of many species.

With the closing of the Tethys Ocean, Australia drifted with the continental flow away from Antarctica into warmer seas, while the corals migrated eastward to embrace the newly arrived land. Both went with the flow, whether of rock or water, continents or colonies.

Whole communities of sea creatures may be found rafting along. A flood-uprooted log may float out from the coast of Papua New Guinea, the Philippines, or the Americas and drift with the currents of the Pacific Ocean. Before long the tree, perhaps with dirt tangled in its roots and leaves still attached to its branches, becomes a focus for life unknown in its former home of the forest. Seaweed takes hold on the branches, and gooseneck barnacles attach themselves to the bark. If the tree passes over a coral reef, reef fishes and their larvae, whether trigger fishes or trevally, are caught up and swept into the open ocean, where they never usually venture. Small damselfishes make a new reef-style habitat in the tangle of branches. When the tree drifts past another reef, the aptly named "drift-log fishes" make a dash for it and find a new reef home.

Continents, colonies, and communities are busy going with the flow. Organisms grow, reproduce, and die, the death of some providing the means of life for others. The drift log is like an ark, a life support system for its communities as they float with it. Planet earth is like a drift log writ large. Earth heads off, its life forms sheltering on it, as it rafts through the sea of space, beholden to solar and planetary forces. "Going with the flow" is a notion that helps in imagining the tensions between ourselves and nature, between what humans can control and what controls us. The task is how best to gain knowledge of the limits set by the natural world and to work out ways of moving appropriately within those limits. Corals provide an exemplar. They take nourishment from the water. They flow with the water and make use of it. They build defenses against it. In death their skeletons contribute to the structure through which water penetrates and around which it must divert itself.

Snorkeling, diving, and swimming are activities to which the term "flow" is particularly appropriate. There is the literal flow of water over skin or wetsuit. There is the nature of the activity itself—the moving "here and nowness" of it, and that "otherworldness" that also takes the individual beyond the "here and now." The sensation that people feel when they act with total involvement is what the sociologist Mihaly Csikszentmihalyi felicitously terms "flow."[6] The diver, the swimmer, the snorkeller, the surfer, when they "lose themselves" in activity, experience a flowing unity in which the distinction between self and environment blurs. In such activity there is a merging of action and awareness that progresses according to its own internal logic. Time ceases to be relevant; immersion in the present moment is everything. There is a "letting go" of the past and the future. The Australian Olympic swimmer Murray Rose says that the principal quality required of the competitive swimmer is a "feel for water." Immersed, attuned to the water pressure along his body from his feet to his face, hands cupping the water under his body as he pulled each stroke through, Rose swam into the record books in the 1956 Melbourne Olympics, in conditions he liked best in competition—at night, under floodlights.

Swimming always seemed, Rose said, like "an adventure into a different world."[7]

Going with the flow. Effortless action. Flowing in harmony with natural rhythms. Sounds a bit Taoist. Taoist philosophy, as given in the *Tao Te Ching*, has its origins in China as long ago as perhaps 600 B.C. Water is a principal image used to illustrate the nature of *Tao*, the Way. *Tao* is the nameless beginning of things, the universal principle underlying everything, the supreme ultimate pattern, and the principle of growth. *Tao* is like an immense boat that drifts freely and irresistibly according to its own will. Move with it, observe the natural world, and learn to follow nature's way or *Tao*.[8] Better to ride along with it than to oppose it, just as, in the martial arts, the strongest person is the one who makes use of the enemy's strength.

The *Tao Te Ching*, in one of its modern guises, is a book that speaks strongly to Westerners today. On one dive trip I read it in the version published by a science fiction writer I greatly admire, Ursula K. Le Guin. She has long loved the book, ever since she discovered a copy owned by her father, the eminent anthropologist Alfred Kroeber. One day she found her father marking up passages he wanted read at his funeral, and she became interested in this book that could inspire and impel a man like him. In collaboration with a scholar of Chinese, J. P. Seaton, she published her own version of the classic text. What delights her is the way the text speaks to people everywhere, as if it had been written yesterday instead of some two and a half thousand years ago. For example, the people living in the Taoist Utopia take the attitude that it is all right to have and use machines, providing they don't come to demand more of you than you get from them. Taoists, she says, don't surrender power to their creations.[9] Little matter that the poems are now read in the thoroughly modern context of computers and mobile phones. Le Guin creates a text for readers today who are looking for a voice that speaks to them about things that matter. It spoke to me on the Coral Sea.

The magnificent Great Barrier Reef has grown from scum. Coral larvae settle, grow, and bud into colonies and then reproduce, die, and eventually become a reef 1,550 miles long and a hazard to shipping. Travelers go to the reef as tourists and take books on spirituality. They dine at the knowledge table set out for them in the late twentieth century, a snippet of marine biology from here and a morsel of ancient wisdom from whatever tradition. There are connections to be made, not all of them linear.

Naturally, the Chinese ancients thought about water differently. The water of the *Tao Te Ching* is not the Coral Sea but the metaphorical waters of life and the practical everyday aspect of the water to be found in rivers and puddles and irrigation ditches.[10] Still, to a reef traveler today, the words of Ursula K. Le Guin's version of the *Tao Te Ching* resonate directly:

*What's softest in the world*
*rushes and runs*
*over what's hardest in the world*
*The immaterial*
*enters*
*the impenetrable.*[11]

I float on the sea. Below me a great solid structure has grown from larvae settling into the ooze and growing toward the light. It has grown both because of water and in spite of the flow of water. It has grown by using the water and in growing has changed the waters through it and around it. Coral reefs influence the chemical balance in the earth's oceans. Reefs serve as a protection against the power of waves. Some connections are logical; some are not. Facts about the curious habits of reef animals get us so far, but there is more:

> *Nothing in the world*
> *is as soft, as weak, as water:*
> *nothing else can wear away*
> *the hard, the strong*
> *and remain unaltered.*[12]

Water flows around objects in its way, along the path of least resistance. In so doing it becomes a force in itself. The term for "actionless action" is *wu-wei*, going with the flow. The aim is to flow with the world in such a way that from this flow comes the power not only to survive but also to effect change. Life on coral reefs works like this, up to a point, apart from the fact that each organism is part of a food chain for others. The Taoist principle of *wu-wei* involves an element of letting other creatures alone, both human and nonhuman, so that they can develop in their own ways. The Taoist on the reef strives to see things from even the corals' point of view. There is more to the world than the everyday, and it can be imagined. Deep down at the polyp level of existence, life glugs and plops and gurgles on, oblivious to aerial commotions.

Ursula K. Le Guin tells why the *Tao Te Ching* and its putative author interest her so much: "Lao-tzu is tough minded. He is tender minded. He is never, under any circumstances, squashy minded."[13] (And perhaps he did not exist or is not one author but many.) Of relevance to the diver is the image of the *Tao* as the immense boat that drifts freely and irresistibly, and the Taoist understanding of breath as more than breath, as life force or ch'i that is best conserved by going easy from the start, so that it gathers power. Once your store of ch'i, given to you at birth, is exhausted, you die. Save the breath, go with the flow, look after your life, and follow the way of virtue.[14] Good advice for surviving on land as well as underwater. What the *Tao Te Ching* offers today's readers is a set of metaphors and myths that at first glance are far removed from reef sciences yet on closer scrutiny are intertwined with them.

Today's Great Barrier Reef is surprisingly young. Eighteen thousand years ago when the earth was colder, the sea level was lower than today. The land that emerged from the ocean united Australia with New Guinea and Tasmania in one continent, and the east coast of what is now Queensland reached out to the edge of the present Continental Shelf. At the sea's edge stood a limestone ridge some 165 feet high, the remnant of previous barrier reefs, while offshore lay only a narrow fringing strip of coral reef.[15]

The present Great Barrier Reef began to form some 9,000 years ago when the seas rose again, together with the sinking of the coastal plain.[16] As soon as enough water covered the land, corals began to grow from the ground beneath.[17] At first, with the rising waters, the reefs grew vertically toward the surface. Then as the corals reached air they branched and developed their nooks and crannies, their wide range of hidden spaces that reef animals claim for territory. Corals grow toward the light and, as they are broken and die, their rubble is swirled around and accumulates as sediment on the ocean floor. It becomes covered in coralline algae, tiny plants that grow in mats and form a rose-red crust. New corals grow on the debris of the old. Larvae of clams and sponges drift in and settle to join with algae in helping create a reef. Sponges grow from the seabed like the turrets of a castle. Called the "bafflers" of the reef, they slow the water motion and provide quiet places for sediments to settle. After the bafflers come the binders, the organisms that live in the ooze and bind it into something more solid.

The stages in the creation of an island are broadly similar. At first all that will be visible above the waterline will be outcrops of dead coral rock. Wind and waves will combine to destroy the point of the reef, but in transferring sand and debris to the rear they create a sand flat. Soon beach rock will form, and the sandy area will grow in size and stabilize above the high-water mark. The first stages in the creation of a coral island are complete—for the moment, until a cyclone comes and washes it all away—and the island-building process starts over again.

Birds stop by, bringing plant seeds in their plumage and their digestive tracts. Grasses germinate first, and next on the Great Barrier Reef will come the low deep-rooted Boerhaavia shrubs and the argusia bushes. Birds begin building their nests as soon as the sand cay rises above the high-water mark. Their droppings bring phosphates and other minerals. Guano is formed in large quantities. As the island grows in size, brackish fresh water will collect underneath it. The soil is enriched with dead leaves and the bodies of dead birds. Straggly dusk-colored casuarinas, lush pisonias, and coconut palms take hold as the island grows large enough to nourish them.

A mature coral reef is poised in the balance of the forces of erosion and the forces of calcification. Corals die and pieces break off and fall in the sediment. This sandy detritus in turn serves to fill holes in the coral structures. The framework rises. Worms, crabs, and bivalves seek shelter within and on the surface of the coral. They bore into it and destroy it in part. These are the eroders. Reef corals adapt to inhabitants both within and without in a most generous and accommodating way, creating and sustaining a rich variety of life. Each reef has a similar pattern of large-scale growth. At the end of the reef-building process, there will be broadly similar habitats on each reef, no matter how far away from one another the reefs are.

Coral reef systems are often growing, adding structure to a reef over millions of years, until conditions change and boring and eroding organisms and forces get the upper hand. Life and death on coral reefs involve a shift between accretion and erosion, recruitment of new life to the reef, and its mortality, the changing ratios of hard corals to algae, the recovery and the degradation of the reef.[18] Some see in this a harmony or a balance of nature. Others think differently. American ecologist Daniel Botkin argues for an ecology of instability. Life forms have changed climate and the composition of the planet long before today. Bacteria did it first some 3 billion years ago when they began to produce oxygen. Biological changes have led to global changes in the environment, which in turn have led to new opportunities for biological evolution. This long-term process of change has occurred through the history of life on earth, in an unfolding one-way story.[19]

Daniel Botkin takes the step from ecology to the hidden dimension of myth and metaphor. He says that all talk about the *balance* or the *integrity* of nature invokes metaphor. Instead of assuming stability, Botkin finds it just as easy to assume inherent instability in nature. Discord is just as likely as harmony.

Another way of looking at it is to analyze the structure of tales people tell about coral reefs. There seem to be four overlapping layers of story. The first level is the recital of an issue: here is a coral reef; there are the problems it faces. The second level is the level of cause and effect: what chemical and biological factors contribute to its existence. The third level is the structure and supportive cosmology of the reef, by which I mean the theories of biological and geological evolution. The fourth level is the dimension of metaphor and myth.[20] Balance, imbalance; harmony, discord; integrity, the relevance or irrelevance of a concept of human probity as applied to nature—these terms inevitably enter discussions of the ecology of the reef and are all ways of knowing the reef.

Long before life took off on land there were reefs in the sea. Reefs were different then. Four hundred and fifty million years ago the reef builders were not the corals of today but sponges (the stromatoporoids), the sea mosses (the bryozoans), and other classes of organisms, many now extinct. The earlier forms of coral, the tabulate and rugose corals, were like modern corals in that they formed hard reef-constructing skeletons that provided niches for other forms of reef life, like soft-bodied creatures that left no fossil record. Remnants of these ancient reefs can be found, now well above the sea, in the Canning Desert and the Gogo region of northwestern Australia or, closer to the present-day Great Barrier Reef, the fossil reefs of Charters Towers in Central Queensland. At the Gogo fossil field the reefscape of 450 million years ago is landscape now. Finding a reef high and dry and stranded in outback Australia brings powerfully to mind the shifts back and forth in geological history from former seas to present land, from land now to future seas. Long before humans were there to see it, the Gogo reefs grew and died and left their structures in the shape of inland atolls and fossil fishes. Earth's timescape is here read in the reefscape. Life evolved in the oceans, and reefs existed then in full, if different, complexity. Ice ages came and went, continents collided, reefs eroded when exposed to air in rock uplift, then grew again as ocean floors subsided. Reefs coped with change and in turn they helped create change, changes to ocean currents and chemistry, even, perhaps, changes to climate.

Four hundred million years ago armor-plated fishes, the placoderms, swam along the sandy floor of oceans that once covered much of Australia. Placoderms dominated life in the waters for some 60 million years before becoming extinct. Geologist and placoderm expert John Long says they were one of the most successful groups of fishes ever known.[21] You can see them on the grassy plains at the Gogo site in northwestern Australia, encased in stone nodules scattered over the sand. Gogo is an eerie place, as if there the sea withdrew from the land and *nothing else changed at all* for 400 million years. The fishes lie near enough to where they died. Whole fishes—the head plates, jaws, back plates, fins, and fragments of skin—all have been preserved in the stone. Today the Gogo placoderm is honored in Western Australia as the state's fossil emblem.

The reef origins of Gogo are celebrated in its place names. Outcrops of rock have names as if they were coral islands. Lloyd Atoll and Wade Knoll are jumbles of rocks rising steeply from the grassy plain. I imagine the ancient ocean as if it lies there still, superimposed on the heat-haze shimmer of the grasses. The years between now and 400 million years ago slide effortlessly away. Lloyd Atoll becomes a coral cay set in a Paleozoic sea. Reefs lie submerged around it—this side for the gentle underwater slope to the lagoon, there the steep fore reef where the open ocean pounds.

Today trees grow in the crevices of fallen rocks. The cliff face is a labyrinth of light and airy caves, roofs open to the sky. In the columns of rock and on the floors of the caves lie fossil sponges, the stromatoporoids, and the curled-up shells of ammonites. Fossil tubes, the relics of worms or tentaculoids, lie mixed with the remains of the ancestors of the crinoids—the feather stars that bring delight today. Fossil shellfishes lie scattered on the ground.

The desert is hot and sweat pours down my skin. Tiny bees settle along my arms and face, drinking the sweat. It feels horrible. I yearn for the desert sands to go away so that I can leap into the ancient ocean. The reef would look familiar in many ways, but the fishes would be very different. Some placoderms, with their armor plates on head and body, had small armlike appendages in place of fins. Their jaws took up much of the head, and their teeth worked on a pulley system—bite and pull. They came in all sizes, to 26 feet long. The placoderms had neither the air bladder of the contemporaneous ray-finned fishes nor the lungs of lungfishes, both organs that help fishes with buoyancy control. Lacking an air bladder, placoderms stayed near the lagoon floor.[22] All Gogo fishes had well-developed lateral-line sensory systems, the sense that tells them where they are in relation to other fishes around them.

I'd get the sense, if I were diving in these ancient seas, that these are fishes and that this is a reef. The ray-finned fishes would seem a little more familiar than the placoderms, having scales instead of armor plates, with a look of the moray eel to them. A third group of fishes, the lungfishes or fishes that can breathe on land, survive today as freshwater fishes in Queensland. The real shock would be with life on land—some insects, a few amphibians crawling from the ooze, and some low plants. I'd feel more at home underwater with the fishes and the corals; the land would look most unpromising as the site for future civilizations.

The placoderms disappeared in a mass extinction event about the time of these ancient reef builders, about 360 million years ago.[23] For whatever reason the waters flowed away from them, leaving the Gogo fossils high and dry and stranded. On land the insects and the plants continued to evolve, and in the sea species like the lungfishes and the ray-finned fishes survived whatever it was that forced some species, such as the formerly successful placoderms, into extinction.

The time-traveling scuba diver now slips forward in time some 200 million years. New forms of coral life have evolved. Single-celled plants have taken shelter inside the tissue of coral polyps and now provide energy and color for their hosts. The single-celled algae, the zooxanthellae, symbiotic dinoflagellates have also entered into partnership with other forms of underwater life, helping create the striking blue and purple colors in the fleshy mantles of giant clams. Corals are not alone in having persuaded plant life to come inside to boost their energy intake.

Closer to the present day, a dive on a reef of some 50 million years ago would point up the difference between land and sea even more strikingly. Underwater, the casual recreational diver would feel very much at home. The ocean at Monte Bolca in north Italy some 50 million years ago sheltered corals and reef fishes very much like those on the Great Barrier Reef today. Fish grazers have evolved and are present in the records at Monte Bolca—the herbivorous fishes that eat algae and reef debris like fragments of coral.[24] Fishes we'd recognize today would swim into view, the Moorish idol *Zanclus* and the batfish *Platax*. Reef fishes evolved rapidly between 70 million and 50 million years ago and since then have stayed much the same. Some 227 species of fishes have been found at Monte Bolca. Something remarkable happened there about 50 million years ago, and everything in the ancient, shallow, warm sea died at the same time. The fishes fell into the calcareous sediment at the bottom where they were preserved, some with fine detail of the scales and the fleshy parts of the body. They show no evidence of attack by scavengers, and the reason is not hard to find. The crabs died along with the fishes. Everything died and sank into the ooze at the bottom. Perhaps a toxic algal bloom swept in and quickly killed everything.

Once again, the story on land is far different. There are trees now in this region of north Italy of 50 million years ago, but the forest is archaic and the land animals totally unfamiliar. The bats would be similar, for example, *Icdaronycteris*—but that would be about it. Perissodactyls (quadruped mammals, with toes on their hooves) such as *Hyracotherium*, the ancestor of the modern horse, could well be galloping around, but they'd be the size of a fox, and our time traveler might not fancy the look of their sharp teeth.[25] Small ratlike mammals scurrying underfoot mark the beginning of the great burst of mammalian evolution. Now rapid action on the evolutionary front will pass to the land, both with the creation of new species and their extinctions. Reef fishes just kept on keeping on. Corals and reef fishes evolve together. They have a good working relationship.

The long history of reef life on earth has witnessed many episodes of reef creation and reef destruction. First single-celled organisms like algae helped build stromatolite reefs, as sediments accumulated and hardened near them in warm, shallow, salty pools. Living stromatalites still exist at Shark Bay in Western Australia, rare remnants of once widespread reefs. The living and the inorganic, the biological and the geological, come together in wondrous partnership. Carbon in the air and ocean cycles into the skeletons of sponges, clams, sea mosses, and corals. The hard structure of the reef is created, then destroyed and replaced, as reef builders evolve and as their activities somehow survive the catastrophes of mass extinctions.[26]

Four hundred million years ago the coral reefs of Gogo flourished and then died out. It took another 200 million years or so before reefs returned but in a different form. Another 200 million years to 1998, and the year this book opens with the story of how the world's reefs face a new danger in widespread coral bleaching. At what point will this event fit into a 200-million-year sweep through time? Will it mark the beginning of the end for corals? Or is it but a blip in a steady-as-she-goes situation—steady-as-she-goes, that is, until the next sharp swift mass extinction event, like the death of the placoderms or of the dinosaurs? What will the Great Barrier Reef be like 200 million years from now? Will there be another reef-building event, of an as yet unknown form of reef? The trend has been from one reef-building event to another to another. Humans build artificial reefs from old tires and concrete blocks. Nature, one hopes, will do better.

# When the Reef Was Ours

What was it like on the reef before the tourists came?

One year my daughter Nora returned from a holiday at Port Douglas with a gift for me. It was a trochus shell, stripped back to its silver iridescent base. The common name for trochus is "top shell," for the simple reason that it is top shaped, if you can imagine a spinning top swirled round with looped silver whirls of soft-serve ice cream. "I bought it from some hippies in the Daintree forest," said Nora. "They said they found it like this on the beach. It's natural." Both of us looked at the shiny shell and burst out laughing. One thing was obvious—the shell was not in a state of nature. It had been stripped of the green slimy accretions that the living mollusk acquires on its shell as camouflage. An unglamorous top layer of outer shell had also been removed. No sensible mollusk is going to heave itself around the sea floor glittering in this look-at-me iridescent glory. This was a trochus shell that had met an unnatural end. Perhaps it was taken live from the reef subtidal zone, buried in sand for ants to strip out the flesh, and ended in some kind of acid bath for the high-gloss final state. Whatever it was, you'd have to say it had been immersed in something *chemical*, a loaded term for those in the nature business. Still, the shell was "natural" in the sense that it was something that once had been alive.

Between 1912 and about 1960, trochus shells were fished in their tons within the Great Barrier Reef. They were fished for the beauty of their shells, not for their flesh. In the Daintree forest, Nora bought one trochus shell at a price unknown to me for it was a present; but I bet the price would have stunned the pearl shellers of the 1930s. At last the trochus has come into its own, in value-added terms, stripped and polished in a local cottage industry in Australia. Decoratively dressed locals with a relaxed attitude sell a trochus shell and an improbable story along with it and to the tourist this contributes to the pleasure. Nora didn't have to believe everything she was told—and she didn't. I like my trochus shell. And unlike other shells, rare and in danger of getting wiped out by this kind of tourist trade, trochus populations are pretty robust.

The iridescence comes from a substance secreted by the mollusk as it lines its shell. Sensible of it, allowing a smooth passage for its soft body. The secretion is laid down in thin films called *nacre*. Fine grooves on the surface create the iridescence as white light suffers interference when reflected from different levels in the groove. The shells glow with light and rainbow color. On my shell, at the base—on the part from which the living animal's foot protruded—there was a row of eight circular grooves, like growth rings on a tree. These rings swirled round the opening and deep into the interior of the shell, wrought by a master engraver, the mollusk itself, which excretes its shell around it as it grows.

My shining trochus shell sits near the computer and prompts glittering memories of reef places. Conversely, it serves as a reminder of the harshness of much reef history, of the worst excesses of the shell fisheries from the nineteenth century on. Not so long ago trochus shelling, along with pearling and the taking of the sea cucumber or *bêche de mer*, were important fishery industries along the Great Barrier Reef and to the north and west of Australia. Today's tourist paradise was once a workers' reef. Trochus and pearl shell, known also as "mother of pearl," were harvested in their tons using largely Aboriginal and Islander labor. The shells were exported for manufacture into buttons, millions of them, and for ornamental inlays in wood. The pearl shell industry started in the 1860s, and the trochus shell industry began about 1912. Both lasted until the early 1960s, with a break during the war years. Some trochus and pearl shell are taken today, still for buttons, and they have found new use ground down as a component of paints and lacquers, but the industry is not nearly as important as once it was.

Pearl shell grows below 40 feet depth, from Torres Strait to the area around Cooktown on the Great Barrier Reef as well as on the north and northwest coasts of Australia. The shells occur joined together in pairs, and grow loose on the sandy floor of the sea. They look like scallops, only larger, weighing up to 2.2 pounds. The larger the shell, the more valuable it is. In the early days of the industry, shells could be found weighing 7 to 9 pounds each. Sometimes the pearl shell will form an irregularly shaped pearl—in about one shell in a thousand. These pearls, however, were not the principal object of the search. While the diver did usually keep any pearls, as a perk in a highly dangerous profession, the shell was the more valuable commodity.

*Bêche de mer*, also called *trepang*, or sea cucumbers, are sausage-shaped holothurians that lie on the sandy floor of reef lagoons. They are mild equable creatures that like nothing more than filtering sand through their gut to extract nourishment, in one end and out the other. Sand is not the only thing that goes in and out their rotund leathery bodies. The tiny pearl fish has made a home in the holothurian, where it feeds on the gonads or sex cells of its host. What a tidy symbiotic relationship these two reef creatures have—the holothurian gut provides the pearl fish with some tasty sex cells, which the host must have in excess; the fish in return provides an internal gut-cleaning service. When the pearl fish wants to leave, it simply exits through the anus, swimming backward. The forces of natural selection have not endowed the holothurian with eyes, so it cannot look back to check if there is a fish swimming backward out of its nether end. The gut of the holothurian is capacious enough to take on this pearl fish inhabitant. Indeed, sometimes it seems as if holothurians are virtually all gut. A favorite trick of the professional nature photographer is to pick the creature up roughly and take shots of what typically happens. The creature expels large spaghettilike strands of its guts, as a defense mechanism. Later it may recover from this inside-outside reversal and grow new internal plumbing, or it may not.

Back in the heyday of the industry, the *bêche de mer* were exported as food to Asia, usually dried. When used in cooking, the sea cucumbers are rehydrated, sliced, and stir fried with ginger, Chinese style, or are used to make soup of a jellylike consistency. The sea cucumber is reputed to have aphrodisiac qualities. Perhaps there's something in it, given that the creature seems to have so many sex cells to spare.

The *bêche de mer* fisheries go back some 200 years before the European occupation of Australia. The early fishers were Malays, also known as Macassans, from the Indonesian islands, and it seems they regularly sailed south in their sailing ships or *praus* to harvest, preserve, and take their *trepang* home. In the nineteenth century, when both shelling and *bêche de mer* fisheries became established in the Australian colonies, it was largely with Aboriginal and Pacific Islander labor. This was the era of "blackbirding," in which

Pacific Islanders were "traded" to Australia for the sandalwood and sugar industries as well as the fisheries. Queensland Aborigines were taken far from home and forced to work on isolated reefs and islands. Abuses were so appalling that the government was forced to pass antislavery legislation, with the Pacific Islanders Protection Acts of 1872 and 1880. In the twentieth century, conditions slowly improved, until the fisheries eventually became more of a joint effort between people of all nations—Malays, Europeans, Japanese, Chinese, and Indonesians. Men who worked in the industry from the early decades of the twentieth century tell many stories of Aboriginal and Asian cooperation in fisheries along the coastal reefs from Cape York to Princess Charlotte Bay.[1]

By the 1930s industrial conditions had improved, at least with an organized system of payment, but the work itself had not changed much. Men still worked from luggers, sailing boats with two masts, called luggers because they were "lugger rigged." In fishing for trochus shell, some 17 men would be cramped into a boat only 33 feet long. After a month-long fishing trip, a trochus lugger might be loaded with 10 tons of shells. At nearly 2,000 shells to the ton, the shells were packed into all available space, including under the bunks. Men lived for weeks at a time in vermin-infested quarters, with the stench of rotting shellfish rising from imperfectly cleaned shells. When the luggers came back to port, they would sometimes be submerged to kill the cockroaches.[2]

Trochus was found on the breaker side of reefs, while the *bêche de mer* were found on the leeward side. The two were harvested separately because they were cleaned differently. A boiler was set up on board for trochus and a smokehouse for *bêche de mer*. Each day men would row from the lugger, four to a dinghy, to find a suitable reef. They would then swim and dive a few fathoms to pick up either trochus or *bêche de mer*. This was called "swimming-diving" to distinguish it from diving proper, which involved the use of a helmet and an air hose. "Swimming-diving" was the less dangerous activity, and the men simply held their breath for as long as they could. The goggles were made by the swimmer-divers themselves, from tortoise shell and glass. There were no snorkels in this period. After 12 hours of hard work in the water, they rowed back to the lugger and often worked far into the night processing the catch further. The first outboard motor boats on the reef were made, according to local legend, by returned World War I servicemen, by lashing a Ford engine to the back of a dinghy.

Swimming-diving for trochus involved first of all finding the shells. Skilled trackers followed a trail of sand they called "trochus shit" across the sand to where the shells lay hidden.[3] A skilled diver could pick up six shells in one hand—one shell on the end of each finger and one in the palm—all in one dive while holding his breath.[4] *Bêche de mer* were harder to pick up. As one Islander swimmer-diver described it: "Pick up *bêche de mer* on top end. Hard to pick up in middle. They made it hard for you to pick them up. Barefoot. Now with flippers you can go down quickly. Before war, no flippers. Only frogmen used them in the war." The processing stage was more hard work: "Boil up. Almost like jelly. If you eat it, you can get sick. If you cure it, it's lovely. More like a jelly. Hard like jelly. Have to strip it, gut it, then boil it. Put in a bamboo pin. Split it, stick it up, and smoke it. Lots of work in it. After it's cured, stack it into a box. Look after it well. If a drop of water falls on one, it spoils the lot."[5] Smokehouses might be set up on islands, and Aboriginal and Islander women were often brought in to do this work.

In diving for pearl shell a diver might wear a "full-dress" diving suit, with copper helmet, lead corselet and boots, canvas trousers, and gloves. These are the diving suits of the early Australian feature movies such as *Typhoon Treasure* (1938). In these movies the diver heroes searched for pearls while fending off thieves, who fought dirty by interfering with the divers' air hoses. In real life, since the pearl shellers weren't after pearls but pearl shell, the horrors they faced were more the natural horrors of sharks or the bends, the pain caused by diving at depth without adequate safeguards. Another staple in the fictional representation of the diver is the "foot in the giant clam" story. This was a highly unlikely scenario for the pearl sheller, who did not dive among corals where the large clams are to be found. Full-dress diving suits were not used in shallow waters near coral because coral might pierce the air hose.

Full-dress diving equipment was heavy. The copper helmet weighed 110 pounds; lead-soled boots were 26 pounds each. In addition, there was 24 pounds in the lead corselet on chest and back. The canvas suit was rubberized and stiff. Underneath, hot flannel underwear and a woolen cap were essential, even in tropical waters, to protect the diver's skin from the punishing pressure of water on canvas. The diver was reliant on his "tender" on the lugger above. The tender operated or supervised the hand or kerosene-powered air pump and controlled the rope lifeline to his diver. Communication between air and the depths of the water was through tugs and shakes on the communication cord or lifeline. After half an hour on the sea floor at 200 feet, the diver had to "stage" his return to the surface, waiting for an hour 30 feet from the surface of the water to avoid getting the "bends," caused by too rapid a change in pressure during ascent. The end product of this work might be a bag with just six pearl shell pairs in it. At the 300-foot depths of the Darnley Deeps, off Darnley Island in Torres Strait, notorious for its dangerous underwater caverns, each diver was supposed to dive only once a day. Stories are told of the ghosts of the Darnley Deeps, of divers who met ships with all sails set, moving over the bottom of the ocean.[6]

It was hard for the full-dress diver to walk across the seabed carrying the weight of the lead. The act of bending down to pick up shell was hazardous, as water could then enter the helmet. Indeed, in the half-dress sometimes adopted for greater freedom, the helmet and corselet rested on the diver's shoulders alone, without being fixed to the heavy canvas suit. The air pressure kept the water from rising inside the helmet. The diver might "throw the helmet" if he got into trouble, a dangerous act that might result in the bends, if he rose too quickly from too great a depth in order to reach the air above.

The mortality rate for divers was as high as 10 percent. In 1954, when the British frogman Mike Wilson took his new aqualung gear to work as a pearl diver, he was horrified to find how swiftly his career progressed from number three position to number one position as the result of the death and injury of his fellow divers.[7] For men to die in this terrible way, and all for pearl buttons, must have seemed the height of human folly to war veteran Wilson. I, for one, certainly have a greater appreciation for the virtues of the humble plastic button.

Shelling was a nasty stinking business. Owen Mass was a trochus skipper who started shelling after World War II. On his first trip he was feeling good about the amount of shell collected, until the stench assailed him from his "stinkers." A stinker was a shell that had been incompletely cleaned and after a few weeks at sea smelled terrible. To fix the problem, Mass tipped the stinking shells into 4-gallon drums of water. Thousands of maggots rose to the surface.[8]

Yet the men who worked on the luggers in the old days recollect shelling with a certain degree of matter of factness and indeed even pleasurable excitement. Historians and anthropologists have recorded many of these men's stories on tape. The historian Regina Ganter spoke to over a hundred old lugger hands from north Queensland, Torres Strait, and Japan. The anthropologist Athol Chase recorded the stories of old lugger hands who came from the Lockhart River Mission. The Japanese skipper known as Kusima was legendary in his daring and skill at navigation and once took his crew all the way through the reef and out into blue water. In three days they reached the Solomon Islands as intended, the skipper steering by the stars and a compass, no sleep, "no maps, nothing else." Certainly no passports or official permission.

In another story the old lugger hand George Rocky recalled a daring chase through reefs and lagoons near Hinchinbrook Island before World War I. The Japanese skipper outran a government steamer that tried to intercept it. "Right! We raced him now. We took off and ran for that main barrier, reefs everywhere there. The Japanese started to sail up right close to the reefs and that steamer couldn't come up. We were scared! The steamer was afraid of the reefs but our captain knew them." Then when it was over, "we all had a good laugh."[9] No wonder government officialdom tried to bring the fisheries under some kind of control. In the years leading up to World War II it was a worry that the Japanese shellers had better charts of the reef than the Australians.[10] The Aboriginal and Islander lugger workers preferred working for Japanese skippers than for the Europeans. The Japanese treated them better and provided better food.

Food on the luggers varied from the absolutely awful to the excellent. Thomas Lowah recorded his time of swimming-diving on a trochus lugger in the early 1930s, under a skipper he later realized was a bad boss. Lunch was thrown at them as they worked in the water—a quarter of a soda bread with syrup—and they were not allowed any fresh water to drink during the day.[11] Another person who as a child lived on a trochus lugger recalled with great relish the food they ate on board: turtle, dugong, *bêche de mer* for a treat, turtle eggs, coconuts, mangoes, wongai plums, and *numus* (raw fish, Japanese style).[12] "It was chewy. Everything was chewy on board the boat. We all had very good teeth." This person gave the following recipe for turtle tripe, which combines the double whammy of two vulnerable species in one dish. Turtle tripe is turtle intestines. "You slice it up, yards of the stuff, you wash it and clean it. It's all wrinkled up. You get the liver with it. It's green. The fat is green and it's a thick consistency, and you fry it. Then you chop the turtle tripe up and add it and it's chewy." If there was some dugong meat it was roasted, cut into chunks, and added to the turtle tripe. Turtle egg sausage was another favorite that could be made on board. "Remove the turtle's egg sac and take out the mature eggs. Tie the end of the tube and fill it with eggs that have not formed shells." Then "leave the sausage in the sun for two days until it is hard."[13]

Reports from people who worked on the luggers show that even though the work was hard, it had its attractions. In the 1930s, Thomas Lowah began work on the luggers as a 14-year-old cook and trochus fisher. When he returned to the sea after World War II, it was as skipper of a pearling lugger until a bad attack of the bends put him in the hospital. He recorded his pride in his own hard work and his achievements and his delight in traveling to distant parts of the reef.[14] The freedom of the seas, the skills of navigation and sailing, and the exhilaration that freedom brings were important aspects of work that otherwise, to modern eyes, looks relentlessly hard and scarcely worth the end product.

My parents were early tourists to the reef. They went to Heron Island in 1939 for their honeymoon. As far as I can reconstruct from photographs, it was a camping affair. My mother never went camping again. Accommodation was primitive on Heron Island in 1939. No luxury resort. Not much fresh water. I hope the turtle cannery that had been there earlier was gone. My father, Rocky, met my mother, Lucille, in America and brought her home to Sydney to meet his family, Irish immigrants, who were a fairly critical lot. Lucille had one of the first refrigerators in Sydney, and that was considered flash, and a bit soft, too Yankee, and what was wrong with the good old Australian ice chest? My mother brought a pop-up toaster with her, and that was a marvel to behold. What gained the approval of her new family was the corn that popped: no one had ever heard popcorn at full pop before. And she had a waffle iron—*that* was wonderful.

On Heron Island my father, Rocky the veterinarian, loved the exotic wildlife and the fishing. He came back with a superb shell collection, from a place that is now a marine park where shelling is forbidden. My mother, the bacteriologist, must have taken to her tent with some horror at the lack of amenities.

Those were the good old days, in the first half of the twentieth century, before the days of mass tourism. Older residents of north Queensland often talk about those days with a sense of loss. A recurrent theme is nostalgia for the old days "when the reef was ours." Once there was sea and sky and freedom. The resources were there, for the taking, and no one was there to prevent it. Now there are too many other people, too many regulations that are now, regretfully, necessary.

The sense of entitlement, that once the reef was "ours," reflects the assumption of privileged access to the riches of the ocean. The reef was ours because a man could come to this place from elsewhere and work, perhaps first as a deckhand or "deckie" on someone else's boat but soon with hard work he would own his own. Then he could go out, as fisherman, or sheller, or shell-grit bagger (shell-grit was sold for chickens to scratch in), or crocodile shooter, or on charter for mining prospectors or fishing groups. A man could be his own boss and lord of the sea. It was largely a man's world, this reef. It was a do-as-you-please life. If a trochus lugger needed wood, it called in at a convenient mangrove swamp. Men chopped and barked the white mangroves they needed to stoke the boilers. Fresh food was easily gained from the sea. Turtle meat and turtle and tern eggs were taken from islands. A dugong hunt might go out from the lugger. No rules, and if there were, there would be no enforcers. The reef was theirs.

Unsurprisingly, stories from the 1960s record a lament for past plenty, a plenty that was often wasted. There are stories of what today would be regarded as irresponsible pollution but which then was regarded as "tidying up." The lighthouse keeper buried his tins in the mangroves, and the mangroves "fed" on them, allegedly, as if rubbish was good for them. Or in a story about boat skippers with their engines: "You're not supposed to dump diesel. Some of these blokes do it in the creek." It didn't seem to hurt the mangroves, this man reckoned, but he wouldn't eat the oysters.[15] Mangroves, in these stories, seem to be endowed with near-mythical powers of pollution resistance. If only it was true.

For the fishermen who conducted their own small line-fishing operations in the 1960s, their nostalgia is for the time before the commercial fisheries, before the trawlers, before the big boys came from down south, before government regulation in the form of the Great Barrier Reef Marine Park Authority (GBRMPA). Refrigeration on fishing boats came in the mid-1960s and made it possible to catch fish in commercial quantities. Line fishing is fishing by lines, not nets, which snag on coral reefs. A commercial line-fishing boat might have a crew of four men and a couple of dinghies or dories. Each day the men went out in the dories and fished the reef. The fishing boat stayed out a month or as long as it took to fill its refrigerators.

Lionel Wickham began fishing out of Gladstone in 1960:

> I built a boat and I went fishing. . . . I went out to the Swains and fished the Swains for 20 years. It was so beautiful out there. We had beautiful lagoons. Once you were inside the lagoons you had beautiful weather. It was so calm. It was something I loved. I'd love to go back now. . . . It was a really terrific life.[16]

Mike Prior from Yeppoon recalled the Riversong area of the Swain Reefs with the comment: "There was a lot more of everything out there in those days"[17]—including shells. One sheller said that when she started out there were very few shellers around. By "sheller" she meant someone who collects shells for sale for decorative or scientific purposes. She recalled: "We were the only ones in the area. We were continually at it. We were selective. Then the crowd came from down south. Shells are getting roughed up now because of bloody tourist boats. A tourist takes them. It's not policed, the shelling."[18] She looked to the southerners, the outsiders, for the cause of damage to the reef. The southerners were the people who turned coral blocks over to find shells and then didn't turn them back again, causing the death of the exposed sea creatures. The southerners were the people who took more than they should. The term "southerners," Thea Astley once commented, when used by a north Queensland resident could include people who lived as far north as Brisbane—in other words, even fellow Queenslanders.

A sense of plenty, a sense of adventure, a sense of danger, and a sense of comradeship were all part of reef memories of "when the reef was ours." Work, often exceptionally hard work, and pleasure are intertwined in memories of those times.

Once there were reef occupations of trochus and pearl shelling, guano mining, shell-grit bagging, sheep farming. Now there are new professions that in a different sense claim the reef as theirs: the scientist, the reef manager, the tourist operator, the nature photographer.

For coastal Aboriginal people there is the sense that "the reef is ours and always has been," despite the European invasion, despite the tourist "infestation." The European tradition is to imagine a cultural distinction between land and sea, in which people may own property on the land but not in the sea. Aboriginal and Torres Strait Islander cultures do not observe this distinction between land and sea.[19] Land flows into the sea and underneath it, where it becomes sea country. "Sea country," as with "country" more generally, has an immense cultural significance. It relates people to a place of origin, to traditional values, resources, stories, and cultural obligations. The sea as much as the land is the place from which these cultural relationships spring.[20]

Imagining sea space in this way is important in the context of Indigenous claims for sea rights. Wayne Butcher, an Aborigine from Night Island, contributed his views to the consultation process on reef zoning with these words: "We want recognition of our sea rights before zoning is carried out. We are still the owners. We want recognition of ownership of the sea country. If we do get it, it will be one thing. It will be Aboriginal sea country."[21] Restoration of marine rights is in turn tied up with increasing pressures being put on Aboriginal people to generate their own livelihood. Too many trawlers are coming in, they say, without asking Aboriginal people and are harvesting the resources without proper care for the future. What if there is an oil spill from a ship near a remote community? People there are not equipped to deal with oil threats and are too far away from major ports for help to arrive quickly.

Claims for marine tenure are enmeshed in a complex of issues, of which the question of earning a living in the twenty-first century is just one. Other issues are: Who owns the sea? Who gets to utilize the natural resources of the sea? Who profits from harvesting the sea? Who loses out? Questions of justice meet high-tech modern fisheries' practices, and high-tech fisheries' practices too often win at the expense of local people and their environment. The practice of taking too much from the sea has, in many parts of the world, led to overfishing. Collapse of fisheries impacts people who claim, with justice, that their traditional methods and systems of ownership worked better for many thousands of years.

In the nineteenth century, fisheries industries came to the Great Barrier Reef, and Indigenous people who previously hunted and gleaned the resources of the reef for their everyday living were contracted into the work force on the luggers. The luggers came and offered the experience of work and then left. Now the trawlers have come, with their high-tech fishing practices that employ few people and that harvest more intensively and destructively. To the west of Cape York, in the Gulf of Carpentaria, the push is on for comanagement of local resources—with some recent successes as local communities and commercial fisheries act cooperatively to tackle the problems.[22]

If marine tenure is granted to coastal Aboriginal people along the coast of Cape York, it may be a limited marine tenure, as with the people of Croker Island. In July 1998 the Australian federal government recognized native title rights to seas around Croker Island off the north coast of the Northern Territory. Five Aboriginal groups now hold a communal native title over the sea. This means the five clans can hunt and fish and protect their sacred sites. But they did not win the *exclusive rights* for which they argued in court: commercial fishermen, mining companies, and pearl shellers may still operate there without requesting the permission of the traditional owners.

One of the reasons for the judgment is that in common law no one owns the seas, that is, the European legal system operates on the assumption that the sea is something that by its nature cannot be owned. Likewise it is assumed that nobody owns the fish until it is caught. The anthropologist Nonie Sharp has made a special study of the sea rights issue and points to the fact that the two conceptions of marine space are inconsistent. On the one hand, there is the belief that "no one owns the seas"; on the other hand, in many parts of the world people who live by the sea do think of local land and marine space as owned by them in some way. After the Croker Island judgment, there seem to be two systems of rights in Australia. Free access to marine space near land is the legal right of two peoples—the five clans and the general public.[23] This, in effect, means everybody.

There are so many competing, inconsistent, and contradictory claims on the reef, when it is claimed as "ours." If the reef is seen as continuous with the land (as reef country was once land) and extending out to the horizon, then it is not sea alone but "sea country." Indigenous people, in making claims for marine rights, are challenging assumptions others make about the boundaries between land and sea, about what constitutes a place that can be owned. Owning a place is partly possession of tradition, laws, and stories that may be read into the place. It will be a place that has boundaries in geography but not boundaries in the imagination. It may be as "bounded" as the sea "from here to the horizon" and equally unbounded, as the sea is free to move.

As long as Aboriginal people have been living near the Great Barrier Reef, they have traveled to offshore islands and reefs. Long before white contact, Cape York was occupied by several hundred kin groups of Aboriginals, each ranging in size from 10 to 50 people. Each person recognized a homeland or country inherited through the father's line.[24] Along the northeastern coast of Australia, at the Nesbit River area, the homeland countries are divided in roughly parallel strips all the way from the mountains down not only to the coast but also extending beyond to the offshore islands, reefs, and sandbars. The homeland country flows from the mountains to beyond the shores and over and under the sea, from dry land to water's edge and beyond to the reefs and islands. There are story places here in which rocks and reefs are located in the context of actions of the mythological beings that first came to this place. The line between sea and sand does not mark an absolute difference—this is land, this is sea—as it does to someone of my cultural preconceptions. To step from land into sea, the way I feel it, is to enter a different place. Not so according to the people who first came to this place and settled there. They came to know the relation between land and sea in ways other than my ways.

I step into the sea and find myself in water. I explain waves by the action of winds and offshore reefs. Yet I also find the sea has powerful resonance beyond this. And I am attracted to stories about the sea told by the Yanyuwa people of the southwest Gulf of Carpentaria. They know the sea as Spirit Ancestor. They see the sea as masculine, the waves as feminine. The waves at one particular place are creations of the sea snake. The crests of the waves are her back, and the fine sea spray is spray from her mouth. The currents found around the islands show the powers of the Spirit Ancestors, and at some places these paths cross. The sea is a place of powers both visible and invisible. The sea is a powerfully attractive place, while at the same time it is a place of danger.[25] Stories of the Saltwater people of the coastline of northern Australia give the traveler a glimpse of what the world of the sea is like for them and a sense that our own stories are rather prosaic and often lack a mythic dimension.

Anthropologist Athol Chase and linguist Peter Sutton have written the history of hunter-gatherers on Cape York Peninsula. The sea-country words they translate give insight into another way of understanding the relationship between people, sea, land, and history. There are so many words from this place to describe the relationship of the land to the sea. The Nesbit River people have words for the zones between sea and beach, such as "deep sea water": *kuytu kulu*; "sea water shallow enough to stand in": *kuytu atya*; "knee-deep sea water": *kuytu nganta*; "sea water edge": *ngaluna*; "intertidal beach zone": *malnkan*; "zone where water has deposited rubbish": *tuyinu palnpana*; "raised sandy area above tidal influence where camps can be made": *yi'an*; "wind-sheared scrub of the foredune area": *malata*. There are words for "the tide-rippled beach," for "wind-blown sand," and for "dirty sand."[26]

I read those sea-country words and reflect that, yes, I have seen that zone where the water deposits rubbish. Yes, I have seen the wind-sheared scrub along the beachfront. But my own language has no words for these special places that come into being as the beach changes daily and as the winds and tides come and go. At Cape Melville, a windy place, the Indigenous word for waves, *waya-ulpa*, is rendered in English as "wind-heaps."[27] A wave is a wind-heap of water.

It makes sense to someone of my cultural preconceptions to say that no one owns the sea because water flows and has no internal boundaries. The sea is both "ours" and it is "not ours." If in common law "no one owns the sea," it is assumed to be everybody's. If it is everybody's, it is not "theirs" exclusively. Freedom is an important element in nostalgia, but the more the concept of freedom of the seas is examined, the more complex it becomes. The more that becomes known about customary marine tenure, the more tenuous the notion is of the "freedom of the seas." Seventy-four claims had been made by 1999 for native title over the sea off Queensland, including parts of the Great Barrier Reef and Torres Strait. The National Native Title Tribunal has accepted 56 claims for mediation.

The Queensland Commercial Fishermen's Organization is talking about the potential loss of the commercial fishermen's rights to catch fish.[28] Once the reef was "ours," say the commercial fishermen. They argue that successful claims for native title will take that right away. In particular, in 1999 a case with repercussions for the future of commercial fishing in Torres Strait came before the courts, when three Murray Islanders were charged with armed robbery of fish. Armed with a crayfishing spear, they allegedly demanded and took the catch from a group of professional fishermen who worked out of Cooktown, outside the strait. The judge, who raised questions about the legal rights to the fish, questions that are yet to be resolved, ordered a stay in the proceedings. Islanders are now proposing a ban on commercial fishing from fishermen outside the strait.[29]

When he was the executive chairman of the Great Barrier Reef Marine Park Authority, Richard Kenchington had a practical approach to nostalgia. He commented:

> The problem is that over the last 20 to 30 years as usage increased, there has been less opportunity for people to do exactly what they want. We have groups of people who are coming to grips with the fact that their opportunities are being restrained because of the reasonable needs of other groups.[30]

Three hundred years ago Indigenous people could and did live sustainably with their reef environment. Both the natural and the cultural environments have now changed. Commercial fisheries claimed their rights to fish the cod of the North Sea to the point of collapse of the fisheries, and tragedies like this impel the need to understand and manage reef fisheries better. New demands on fisheries are emerging, with as yet unknown impacts. Live coral reef fish, from the Great Barrier Reef and Pacific Island nations, are airfreighted to Asia for restaurants, where they fetch large prices as status symbols. In 1995 the live fish trade brought 25,000 tons of live coral reef fishes into Southeast Asia. In 1996 a live groper sold in Hong Kong for $10,256. The corporations that engage in the live fish trade are economically larger than many Pacific Island nations.[31]

Once little was known about the science of the reef; now much more is known about the potential for disaster from both natural and human causes. Forget about trying to accommodate all the user groups, says Jeremy Tager of the North Queensland Conservation Council. Instead, go back to the primary need to protect the reef.[32] Maybe nostalgia isn't going to be what it used to be.

# Reefscape with Sea Serpents

*The Great Barrier Reef of Australia*, published over a hundred years ago in 1893, is a classic in the history of reefscape. In this magnificent book, zoologist and photographer William Saville-Kent brought art and science together on equal terms. The illustrations treat reefscape as if it were landscape, with corals, fish, mollusks, and lagoons as elements in composition, in place of the peaks, forests, and rivers of conventional landscape art.

William Saville-Kent combined in one person two attitudes toward the reef that were later to become polarized—the conservationist and the resource utilizer. Saville-Kent believed the reef was bountiful enough to provide something for everybody. In 1893 he worked out that the reef was worth £100,000 annually, and rising, for the young colony of Queensland. He would have been strongly in favor of late-twentieth-century conservation measures, such as the creation of the Great Barrier Reef Marine Park, and he also saw the reef as a place for people to make a living, within limits set by science. Yet he did far more than chronicle entrepreneurial opportunities. His book is testimony to the pleasure to be found in contemplation of the richness, intricacy, and variety of invertebrate life, in the humble but extraordinary lives of creatures without backbones. He was particularly passionate about oysters.

Saville-Kent was a philosophical traveler in a new world, the world of underwater reef life. He studied the habits of pearl oysters, *bêche de mer*, corals, and fish. He was curious about everything, even the question of sea serpents. He wondered if pearls could be cultivated artificially by introducing grit into their shells at a certain stage of their development. On Thursday Island in 1891 he experimented and had some small success. Although he admitted that "the subject of Worms is not, at first sight, one fraught with great possibilities," he proposed harvesting the marine worm *Palola viridis* for export as a Great Barrier Reef delicacy.[1] He suggested (not entirely seriously) that, one day, members of Parliament would sit down to a "diet of worms" as now they sit down to a whitebait banquet. But in all his reports on the potential of reef industries for Queensland, it never occurred to him that over a hundred years later mass tourism would be the biggest money earner of them all. By 1996 tourism on the Great Barrier Reef would be worth $750 million annually to Australia.[2]

William Saville-Kent was born in England in 1845 and died there in 1908. In 1868 he began work in the Zoology Department of the British Museum where his task was to classify and catalog some 70 species of the stony corals (genus *Madrepora*). The corals were dead and white by the time he picked them up out of the packing case in the museum dry store. As he worked on his catalog he dreamed of traveling the world to see living corals on tropical reefs.

In 1872 Saville-Kent was appointed to the staff of the first large public aquarium in Brighton and two years later to Manchester Aquarium.[4] Large aquariums came into existence with the invention of the steam engine, which allowed large volumes of water to be oxygenated and kept in circulation. The aquarium job was a mix of entertainment, science, technology, and commerce, with Saville-Kent a pioneer, clearly delighted both with the pleasures of observing fish behavior and reproduction and the potential for new aquaculture industries. When the Brighton aquarium resident lobster laid some eggs in 1873, Saville-Kent charted stages in the lobster life cycle, from the undifferentiated egg to the highly specialized adult replete with tough exoskeleton, whiskers, and claws. He could see the potential for lobster farming and later in Australia would experiment with the aquaculture of oysters, pearls, and pearl shells. He was also an advocate for the importance of the young science of marine biology and urged the establishment of a marine biological research station both in England before he left and later at Thursday Island. He was far in advance of his time. He would be delighted to see two of his enthusiasms in full flower in north Queensland today—the Australian Institute of Marine Science at Cape Ferguson near Townsville and the Great Barrier Reef Aquarium in Townsville.

Saville-Kent arrived in Australia in 1884, taking up work in Tasmania as inspector of Tasmanian fisheries. In the next three years he traveled extensively and advised on fisheries in New South Wales and Western Australia. Between 1889 and 1890 he was commissioned by the Queensland government to investigate the *bêche de mer* and pearl fisheries of Queensland and to report on conservation measures in response to concerns about overfishing. At long last he was able to realize the ambitions of his daydreaming years at the museum: to see living corals.

Photographs in *The Great Barrier Reef of Australia* show Saville-Kent at work on the reef. He is dressed too formally for comfort. In one image he wears a white suit and a solar helmet. Ankle deep in water, he pays no attention to the fact that his boots are soaked and that his beautiful white trousers are wet through and clinging to his legs. He peers through a huge camera suspended in a square frame across a rock pool. Another image shows him in black coat, pale trousers, stiff white collar, and hat. He sits on a stool in front of his grass hut, surrounded by specimen trays and parts of his camera. A formal portrait shows a bushy-bearded man with a glint of a smile reflected in his eyes. His hair is parted severely down the middle, greased flat, and slicked back at the sides. His collar is high and starched, a white Victorian shirt under a double-breasted coat.[5] Yet despite his stiff clothes he is a David Bellamy look-alike, and it is easy to imagine him thoroughly at home on television today, recounting his zoological enthusiasms.

Saville-Kent was an advocate of sustainable fisheries at a time, over a hundred years ago, when most imagined the sea as an endless source of self-replenishing bounty. For the pearl shell industry he recommended some important conservation measures, which were adopted, if not always enforced—the closure of some pearling beds and the setting of a 6-inch minimum for the size of collectible shells, to allow the beds to recover and continue to produce.[6] He experimented with pearl shell farming and the artificial culture of pearls, such as the small living shells he collected for investigation on Thursday Island.[7] Pearl shells of this size, about three-fifths of an inch across, protect themselves by secreting a cable, or "byssus," of tough green threads to form an anchor to the rocks underneath. Saville-Kent discovered that severing the cable did not harm the organism; the shells ejected the remaining part of the old cable and secreted new green threads, which soon hardened in the water to create tiny anchors to a new surface. By the time the shells are mature, they have lost the byssus anchor and are found loose on the ocean floor.

For an aquarium Saville-Kent used two large clamshells that stood in the garden of Government House on Thursday Island. His tiny pearl shells adapted themselves "with remarkable alacrity" to their new environment and attached themselves to the inner surface of the large clams. The winds of the southeast monsoon helped aerate the water.[8] The next part of the experiment brought disaster, however, as he tried to move them south to Cooktown. When he replaced the water in the tanks with muddy water from the Endeavour River, he inadvertently killed them all. Even so, he judged the experiment a success because it proved that pearl shells could be moved from place to place on the reef. Future prospects for aquaculture looked good. It would be a simple question of taking juvenile pearl oysters from deep water to pools closer to shore, where they would be easier to pick up when mature. He experimented with the artificial culture of pearls, by inserting grit to act as a nucleus around which the new pearl would form.

The photographs in *The Great Barrier Reef of Australia* capture the reef at exceptional moments of low tide and dead-calm weather. Few visitors see the reef like this. As a scientist Saville-Kent was well aware that the flip side of reef beauty on the windward side of the reef, the conditions under which corals grow at their best, was large bare patches of reef rubble on the leeward side, the conditions under which dead coral helps build coral islands. As photographer he framed beauty in such a way that it proved—not deceptive, exactly—but scientifically misleading. Maurice Yonge, the leader of the Great Barrier Reef Expedition of 1928 to 1929 said rather ruefully that, at the time his team left England for Australia, they knew the reef only from Saville-Kent's photographs and these proved "far from typical."[9] The details are so clear that years later scientists are able to visit the precise locations and measure any changes—just what Saville-Kent intended.

The text weaves round the photographs, delicate reefscapes inhabited by a rich and rare assortment of marine life that few British readers could ever hope to see. Photography enriched science and took it beyond the catalog of bleached museum specimens or scientific illustrations by pen and brush. On the reef the corals are exposed for too short a period of time for the artist to record all the details.[10] "Reefscape" is a term Saville-Kent coined to describe his art. He liked to include the distant horizon with sky and clouds, allowing the fore-grounded corals to rise through the smooth ocean waters as if a weird forest of branched fingers.[11] The genus *Madrepora*, over which he labored in his early museum work, is represented by living specimens in their natural place. Of one photograph he says, "At first glance . . . this scene might be interpreted as the deserted battlefield of some titanic Mongolian horde, which had left behind them decapitated heads and grinning skulls as grim trophies of their desperate encounter."[12] But no, it is only the corals of the aptly named Skull Reef. He points the reader toward a "head piece with a low dint on the forehead," while under the right eye of the "face" underneath flows "the unmistakable vestige of an unevaporated tear." In the days before color photography, all he had were tones of sepia, and these cannot give even a vague indication of the brilliant deep purple-blue colors of the clam's mantle.[13] Oh, the frustration he clearly felt at having such inadequate words. Still, the sepia tones allow the intricate patterns of coral growth to emerge clearly. The dead-flat seas, the exceptional low tide—how patiently he must have waited for those rare right moments!

Photographs make possible a kind of time travel, bringing aspects of the past to view in a selective way that records something of the photographer, his time and culture. They offer rich potential for future interpretation. Saville-Kent photographed a group of women on Warrior Island in Torres Strait preparing *bêche de mer* for the smoking process. He showed what he believed to be a happy group of willing workers, adept at a job for which they were well suited, with the horrors of the early days of the industry now long past. One hundred years later Henry Reynolds uses the same photograph in his book, *With the White People*, to illustrate the horrors of the *bêche de mer* industry, still the same nasty stinking business it ever was.[14]

Saville-Kent carved out a living from the new science of zoology and showed its practical importance to government and industry. From the museum study of dead specimens, and the technological artifice of the large public aquarium, he came to the real thing in a place far from home. His photography made the reef seem "realer" than real. He was one of the first voyagers in "hyperreality." Travel to new places was part of his work, though it could be said that the boundary between work and recreation for an enthusiastic zoologist is blurred. Wherever he went, whether to live or to work, he set up menageries: frilled lizards, owls, cockatoos, and parrots lived in cages around him. When living at the Grand Hotel on Thursday Island in 1891 he installed a cage of parrots and cockatoos on the verandah, alongside foul-smelling dead and drying corals and anemones. Ellis Rowan, the artist and botanist, was a fellow hotel guest who understood his enthusiasm. Her room was filled with flowers. Boundaries between life and work did not exist for these two.

Saville-Kent's passionate interests brought him to Australia, but there were other reasons why he may have wanted to start a new life. His biographer, A. J. Harrison, tells of "a turbulent and terrible childhood."[15] At the age of 11 with his 12-year-old sister, Constance, William ran away from his tyrannical father and his abusive stepmother. They were caught and returned home for more punishment. (His story has strong resonances with Charles

Dickens's novel, *Edwin Drood*.) Worse was to follow. When William was 15 years old, his three-year-old half-brother Francis Savill was taken from his bed and brutally murdered. His throat was cut and his body dumped in a privy. Some years later his sister Constance was convicted and jailed for the murder. At the time there were grave doubts about her confession, and even the judge was in tears as he sentenced her. The murder was as much a mystery as was Constance's confession, which few people (including, it would seem, her brother William) accepted as true. After 20 years in jail she was welcomed by William and her other siblings into his family when she migrated to Australia to be with him. Constance Kaye, as she called herself in her new life, lived an exemplary life of devoted social service as a nurse and died in Sydney at the age of 100. Constance and William shared a terrible childhood yet emerged from their ordeal to make valued, if quite different, contributions to Australia.

William Saville-Kent was a pioneer of a new profession—the fisheries expert. His work reached out to more than the zoology of oysters, pearls, and fish, to embrace the meaning of fishing in culture. He was prepared to consider all information that came his way about the Great Barrier Reef and its inhabitants—even the sea serpent. Stories of sea serpents abounded, but he had yet to see one. Sea serpents might, if they existed, pose a threat to the fisheries. He was prepared to look at the evidence and keep an open mind.

Who knows what to believe about the sea? Stories that sound fantastic sometimes turn out to be true. As Robert Johannes says in his delightful book on the fishermen of Palau, *Words of the Lagoon*, many stories of marine animals with habits that once seemed fantastic are now accepted by science. Johannes reports seeing for himself some marvels he previously doubted: the tree-climbing octopus, the triggerfish that builds a large nest for its eggs, the archerfish that spits water in jets to shoot insects from the air.[16] About the sea serpent, however, Johannes is silent.

In the realm of sea lore, fact or fiction, stories of sea serpents have long held a special place. The Great Barrier Reef has its own sea serpent story. The sea serpent with the Aboriginal name the "Moha Moha" was sighted at Sandy Cape (to the south of the reef, admittedly) on Fraser Island in June 1890. The first recorded sighting was by someone known only as Miss Lovell, who described what she had seen in a letter to a friend. Subsequently the letter was published in a London hunting and fishing magazine, *Land and Water*, in January 1891. Miss Lovell reported that one day she saw on the beach near the lighthouse a creature that was unknown to her and to science. She wrote:

> We have had a visit from a monster turtle fish. I send a sketch of it. It let me stand for half an hour within five feet of it. When it tired of me looking at it, it put its long neck and head into the water and swept around seaward, raising its huge dome-shaped body about five feet out of the water, and put its twelve feet of fish-like tail over the dry shore, elevating it at an angle. Then, giving its tail a half-twist, it shot off like a flash of lightning, and I saw its tail in the air a quarter of a mile off, where the steamers anchor. It has either teeth or serrated jaw bones. Native blacks call it the Moha Moha and say they like to eat it, and that it has legs and fingers. . . . I think it must be 30 feet in all.[17]

The *Land and Water* expert suggested it was probably a monster turtle, *Carettochelys*, from the mouth of the Fly River in New Guinea, and the "fair observer" must have been mistaken about the teeth. When another expert, Bernard Heuvelmans, examined this evidence much later in 1965, he ridiculed this species attribution because *Carettochelys* was a small creature, only some 30 inches long. Perhaps, Heuvelmans suggested, the *Land and Water* "expert" made his claim from a sketch, without reading the detail in the description.[18]

Miss Lovell was similarly incensed, for different reasons. She sent a spirited response to the criticism in *Land and Water*, which was published three months later because of the slowness of the post between Australia and England: "You speak of the impossible length of its tail. I beg to state this is a most astounding statement from people who have never seen this monster, half fish, half tortoise." The tail, she said, is "twelve feet long, wedge shaped, and has semi transparent chocolate flukes, about 18 or 20 inches in length. . . . The rest of the tail is a beautiful silvery gray shading into white." She was emphatic: "It is not a turtle, but a monster."[19]

Part of the problem with the account of the sea serpent, according to *Land and Water*, was the nature of the witnesses. Miss Lovell cited, in support others who had seen it some time before, Aboriginal informants who named the creatures from the sketch she made: "Moha Moha. Saucy fellow . . . dangerous turtle." They told her it had attacked their camp eight years ago and knocked over a man with its front feet. *Land and Water* replied disdainfully: "We lay no stress on the evidence of the *blacks*, as all colored races are greatly given to exaggeration and invention." Instead, the London expert showed himself to be greatly given to exaggeration and invention by telling Miss Lovell that what she saw was not one monster but two separate creatures. The Moha Moha was nothing more than a very large turtle clasping a very large fish in its hind limbs.

At this Miss Lovell grew furious. She shot off more letters about what she had seen to Dr. E. P. Ramsay, a curator at the Australian Museum in Sydney, and to the Sydney naturalist A. H. Chisholm, to whom she sent a color sketch. She reported a number of further sightings, once by seven whites and two blacks. "One old sailor thought it was some kind of whale; a lad of nineteen and three girls thought the head was that of a crocodile." One way to find out for sure would be to offer a reward for the animal, dead. It would have to be shot in the head, she suggested, because if shot in the tail it would only be wounded and would sink to the bottom of the sea.

So much for the splendor and rarity of the sea serpent. If you find one, shoot it quick, and stuff it for the museum. You will get, said Miss Lovell, "£100 for the entire animal, £50 for part, and a fair price for the head and neck, sun-dried." No wonder sea serpents stay well hidden.

On the other hand, look at what happened to Miss Lovell. Her letter goes forth to a skeptical editor of *Land and Water* who knows sea serpents do not exist. He helps create a climate in which people will not tell when they think they have seen one, lest they be thought drunk, deluded, or both. The testimony of Indigenous peoples is rejected, yet it is precisely people going about their everyday business who are the most likely to be there, if there is where the extraordinary is about to happen. Perhaps that is why there have been no more reported sightings of the Moha Moha. People have seen it but prefer to keep it to themselves.

What exactly did Miss Lovell see? Her descriptions are very detailed. She saw something quite unusual. It was a creature with a long neck, 4 feet, and a huge domed back that was capable of breathing out of water. Her conclusion was "half fish, half tortoise." But her correspondents wanted it to be one creature or the other (or two in the case of the turtle holding the fish). *Land and Water* adopted a lofty tone: "From a scientific point of view, the existence of a creature combining the characteristics of a fish, together with those of a tortoise, [is] absolutely impossible," and reminded readers that "scientists who have made Natural History and anatomy their life's study" (as Miss Lovell had not) have so far failed to find turtles with teeth.

Heuvelmans, however, with the benefit of 1960s knowledge of fossil fish, was prepared to concede that there could once have been a creature that was half-fish, half-turtle. The fossil record shows plated-plated fish, and the Moha Moha resembled the long-extinct placoderm *Pterichthyodes*.[20] European scientists had scoffed at the idea of a platypus as an egg-laying mammal. To Miss Lovell and her Australian scientific correspondents, the world about them was yielding wonders daily. If Charles Darwin asked his readers to imagine a common ancestor to both apes and humans, if he stirred the popular imagination to create a notion, however inaccurately, of "the missing link," then why not something half-tortoise, half-fish?

Another question asked in the prodigious literature about sea serpents is: Why is it that records of sea serpents exist if sea serpents do not? Is it some kind of "perceptual contagion" perhaps?[21] The term describes the situation where first one person sees something, then another falls victim to suggestion, then another. Expectation is created and the image is already in the mind. It then happens that there are a number of people who report having seen the same thing, like the Moha Moha, even though the person who started the train of "perceptual contagion" may have been mistaken.

Bernard Heuvelmans had an ingenious suggestion to make: that Miss Lovell's drawing resembles the *makara*, the sea monster of Indian legend. He provides a sketch of a creature with the head of an elephant, a fishlike body with a camel hump, four legs like a lion, and a forked tail like a fish. Declared Heuvelmans: "I find it hard to believe that Miss Lovell was not a dotty old maid who had picked up, but not digested, a smattering of paleontology and Brahmin legend."[22] Just as well Miss Lovell was long dead and didn't get to read this version of her story. The testimony of old maids joins the testimony of blacks as unreliable. There are important race and gender issues in sea serpent spotting. Nor does Heuvelmans record how Miss Lovell might have got hold of his sketch of a *makara* at Sandy Bay in order to have her imagination overstimulated by it. He'd have to check the records of the Sandy Bay public library, if such a place existed, in order to pronounce with such assurance.

Rupert Gould is another enthusiast for the sea serpent who has evaluated Miss Lovell's report more sympathetically. Writing in 1969, Gould reckoned that a giant turtlelike creature possibly does exist. With respect to Miss Lovell, he added: " . . . nor do I think her report was quite fairly treated, in scientific or quasi-scientific circles, on its original appearance."[23]

William Saville-Kent treated Miss Lovell fairly. He asked her for more information about her adventure, and he also gathered corroborative evidence from witnesses. He concluded that she did see something but that it may not have been what she thought it was. Although somewhat skeptical, he suggested a scientific name for the creature, *Chelosauria Lovelli*.[24] He included her illustration in his Great Barrier Reef book.

When another "sea monster" was found dead on the west coast of Tasmania in 1962, the *Australian Women's Weekly* ran an article on it.[25] Whatever it was, it had been dead about two years, so there was plenty of room for conjecture. Was it a giant ray, a whale, a Kraken liberated by the Pacific H-bomb tests, or "an eerie thing from perhaps a million years away" that had been released after being first entombed in the Antarctic ice? Everyone had a theory. Perhaps it was "the Blob," a Martian astronaut, or an escaped haggis, fully grown. And to contribute to the debate, the "celebrated old-timer—the Queensland Moha Moha" was produced. The sketch Miss Lovell sent to the Australian Museum in Sydney appeared side by side with photographs of the Tasmanian Blob.

The sea serpent or its lake counterpart, the Loch Ness Monster, has long been part of oceanic folklore. People who have spent their lives at sea have seen strange things. They may be explained, rationally, as giant turtles, giant squid, or the large sunfish or oarfish that bask on the surface of the water. To the sailor near the end of a long watch, they may appear as sea serpents in some lights, some strange oceanic conditions. Exhaustion takes a human toll, and the mind sees visions that have more to do with dreams than with the world that is visible on the external waters. Perhaps also it is the case that the sea, over the millennia of human culture, has accumulated so many associations, symbols, myths, and metaphors that it produces a hallucinogenic effect. Or, if there are sea monsters out there, they may be hidden in the deeps. Huge crabs, tubeworms, and mussels have recently been filmed, and captured, from the deep recesses of the sea, from the "black smokers" or hydrothermal vents above fissures in the ocean floor from which seeps the heat of the inner earth. These forms of life are so weird, with their fluorescence, sulfur-based metabolism, and strange defense mechanisms that they make the sea serpent seem tame. These forms of life exist under such huge pressures that they explode as they are brought to the surface.

Bernard Heuvelmans suggests a new word for the study of hidden forms of life: *cryptozoology*. He seeks an input into zoology from local, Indigenous, explorer, and traveler traditions, sightings, tales, and legends of as yet unverified animals. These are the animals of folklore and personal testimony, not yet the animals of the formal classification in zoology.[26] He seems to have rejected the sea serpent of the Great Barrier Reef. It will take more testimony than Miss Lovell and her informants to convince him, and, alas, further witnesses have not come forward since her time. If only William Saville-Kent had been there to take its photograph.

# The Tears of the Turtle

If reincarnated, I will never come back as a marine turtle—certainly not one of the edible varieties, the green turtle *Chelonia mydas*, for example. No doubt reincarnation does not happen by personal preference, but if I had even a minimal say in the matter I'd say, all right, I'll go back as a sea turtle but only if I'm a male. The female turtle has the worst time of it. Turtles are protected in the Great Barrier Reef Marine Park, but elsewhere to the north and along the coast of Papua New Guinea and Indonesia it is always open season on them. They are taken for their flesh and their shell.[1] As food, the female is preferred for her fat and her eggs, and she is easier to catch when she comes up on land to lay her eggs.

Female turtles begin laying their eggs at age 50 and then come back to lay them every six years for another 50 years. Each clutch is from 80 to 120 eggs. If she is lucky enough to come up on a beach in protected waters, she may find herself under close observation from turtle watchers, who mean her well but are still intrusive of her privacy. Turtle watching is nature's own spectator sport. There's the fun of waiting for the turtles to emerge from the sea on the dark, warm beach or of following the distinctive turtle "tire" tracks up the sands to where she is digging. The turtle digs while people watch and wait quietly so as not to disturb her and send her back to the water before she has laid her eggs. How hard she works at digging, how clever she is at using her hind flippers to scoop out an egg chamber. Then the moment of ultimate voyeurism comes, when she settles to it, extrudes her oviduct and squeezes out egg after egg after egg. When she is judged to be totally engrossed in the laying process, someone will shine a torch on her rear end to illuminate the proceedings. Nights of turtle watching are made of scenes like these. I wouldn't be a female turtle, ever.

The turtle-watching industry is worth about $1 million a year to tourism in the Bundaberg region of Queensland.

The hatchling is imprinted with a peculiar magnetic sense so that it knows the dip and strength of the earth's magnetic field, and this information will later guide the females back to this particular nesting site to lay their eggs. Some 50 years later, having traveled thousands of miles, the females come back when they are ready to nest.

I feel a forlorn kind of comradeship with the female turtle. All the cards in nature's bundle of horrors are stacked against her. The sea is her place of refuge, but she must come out of it onto a beach to lay her eggs, and it is the females that suffer disproportionately in harvesting. The female has traditionally been preferred for food. If a copulating couple is spotted and turtle hunters set off in pursuit, it is the female—heavier, slower, fatter—that is taken, while the faster male speeds off.

The way to avoid being born a female turtle is simple and effective. Unfortunately, the hatchling is not in any position to control it. Work done by Queensland scientists shows that the temperature of the eggs determines the hatchling's sex during incubation.[2] Cool nests produce males. Avoid the warm nest: that's the message for the turtle that does not wish to be female.

Being a turtle hatchling is unenviable, whatever the sex. The first few hours of life are hazardous, as the young break from their soft eggs and struggle upward through the sand. They orient themselves toward the sea by means of the horizon line and waves. No mother love for them. They are well endowed for immediate independent living, but it does not help them in their first few perilous yards of life, when they must reach the sea before being grabbed by a waiting sea bird or a voracious crab. When this scene is shown on film the viewer is expected to take the side of the innocent turtle hatchling against the vicious crab or bird. My view is, better to be taken then, little turtle, better to die very young if you're a green turtle, than to be taken as an adult and killed in nasty ways—upended on the sand and left to die, as happens in places where they are taken for food, or cut up while still alive. Pity the turtles and cherish them, for they too are on the conservationist's list of vulnerable species and in danger of extinction.

The first movie ever made of the turtle laying its eggs was censored in Great Britain. The oviduct descending and squeezing was deemed—obscene? Vulgar? Perhaps unsuitable for family viewing at the end of the newsreel segments, which would also, at this time, have shown the rise of fascism in Europe? *Ocean Oddities* (1929), one of the first movies of the Great Barrier Reef, had parts of the egg-laying sequence cut by the British censor. Filmed by Noel and Kitty Monkman at Green Island, it is in black and white and is nine and a half minutes long. The turtle is shown laboring up the beach, with the Monkmans following. Tall robust Noel and tiny Kitty also labor, with their heavy lights and tripod and 35-

millimeter camera. The turtle settles and digs its egg cavity with its hind flippers. The white fleshy oviduct is extruded, and the eggs are squeezed out. How delighted the Monkmans were to be the first to capture this scene on film. What hard work it was for them, though they didn't get everything right. In the film the narrator at times absentmindedly refers to the egg-laying creature as "he."

When news of the turtle censorship in Britain reached the Sydney papers, great was the merriment. One cartoon showed a stork carrying a tiny turtle, with the caption: "Can it be that the English still believe the stork brings the baby turtles?"[3]

Noel Monkman came to underwater photography from an initial interest in microscopy, as did his colleague Bruce Cummings (b. 1902). The two men were pioneering photographic naturalists. In making their nature documentaries, they had to do everything themselves, from building their own diving equipment and underwater motion cameras to editing their films. Monkman (1901-1969) had earlier been a musician and amateur microscopist. Together with his wife Kitty he lived for six months on a coral island in 1929 while they made their first five reef films. Bruce Cummings was a Cairns radiographer who worked for the radiologist and eminent naturalist Dr. Hugo Flecker. (Flecker had the dubious pleasure of having the deadly box jellyfish *Chironex fleckeri* named in his honor.) Cummings's reef images were shown internationally in the Fox Movietone film, *Wonders of the Sea* (1922). Monkman and Cummings worked together on the sepia-tone feature film *Typhoon Treasure* (1937). Monkman wrote the script and directed the adventure yarn of shipwreck and lost pearls, while Cummings did the photography for the underwater sequences, shooting from a diving bell he made himself.

Monkman made his first film in Sydney in 1922. In *Monkman Microscopics: Strange Monsters in a Drop of Water* he filmed a familiar nature study exercise. Pond water looks clear enough to the eye, but place a drop under the microscope and hidden wonders are revealed—a wild new world of tiny transparent wriggling life, creatures that pass their short lives in everyday places and in hidden ways. The movie opens on the banks of a lily pond, with naturalist Monkman snappily dressed in suit and white hat and carrying a cane. He collects some pond water in a jar. The next cut is to the view through the microscope. "What is this?" asks a surprised male voice-over. "A sea serpent or a Chinese dragon?" No, it is the larva of a mosquito that has just hatched from an egg—enlarged 10,000 times. It is transparent, with its internal parts clearly visible in the strong black-and-white contrast. The heartbeats, and wriggling mites, much smaller still, pulse down its gullet. A swiveling proboscis suctions like a fierce miniature vacuum cleaner. "Look, there they go," comes the voice-over. "No return ticket on this joy ride!"

The film was first shown commercially in 1931 on Fox Movietone News.[4] It was also used, together with some of Monkman's film of the reef, in the American film of Rachel Carson's best-selling book, *The Sea Around Us* (1951). This film won an Oscar in 1952 for best feature documentary. Altogether Monkman made 28 films in partnership with his wife, Kitty, ranging from newsreel clips to longer nature documentaries and feature films (including *King of the Coral Sea* with Chips Rafferty). Monkman ran a cinema on Green Island where he showed his own films exclusively. Bruce Cummings photographed the microscopic life of plankton and made the first film on the life history of the sea eagle.

Monkman and Cummings both took great delight in the reef while at the same time seeing most of the objects of their study in unsentimental terms as food. Monkman's voice-overs sound rather arch to the modern ear, as he describes the quaint habits of the turtle or the mosquito larva. He frequently puts himself in the picture, sometimes doing things that seem rather dreadful today. *Ocean Oddities* is a curiosity of the pioneering days of nature documentaries. Many of the "oddities" of the reef are shown dead or gasping. In this film there are no underwater shots of marine life in its natural habitat. The creatures are out on land and decidedly unhappy about it or dead. The dying stonefish is prodded with hooks and knives to turn it over. As it gasps for air, the voice-over intones, lugubriously: "The mouth is a ghastly cavern." One sequence shows the heart of a turtle held in a human hand, the heart still beating strongly. "Dead three days," announces the voice-over, "but still beating." Can it possibly be true, after three days of—what? But the heart was beating and it was held outside the turtle body though still attached. A dead crown-of-thorns starfish on the sand has a bare human foot descending on it from above, to the words "Hard to put your foot down." A *bêche de mer* has ejected its guts, as it does when threatened, and looks like a sausage skin. It is motionless. "It is able to grow its missing parts again," the voice-over asserts with a confidence the viewer does not share. So far the count is four dead or dying reef creatures.

One final scene has Monkman frying a turtle egg on his camp stove. He first finds the eggs by jabbing into the sand with a grass spear, as the local Aborigines taught him. "If it comes up sticky, dig down." He cracks an egg and fries it. "Only the yolk is good to eat. No matter how much you cook it, the white never sets." The egg has a tough center with a sticky translucent white. The egg-frying scene is a Monkman specialty. In *Island of Turtles* (1958) he fries up a hen's egg and a turtle's egg for comparison, and in *Nests in the Sun* (1956) he fries the egg of a noddy tern.

There is a directness and immediacy to Monkman's documentaries. Today, as in films like *Coral Sea Dreaming*, the viewer has the illusion of being under the water with the photographer. But the coral polyps may be spawning in the aquarium tank, not the open ocean. *Ocean Oddities* portrays a mix of dying, dead, and fried things, and in so doing it brings the viewer close to the realities of life on a coral island, as lived by the Monkmans in the early decades of the twentieth century.

The eight newly hatched green sea turtles are tastefully arranged in a large soup tureen filled with water. They are alive and kept separate from each other by fronds of seaweed placed in between them. Their heads point outward and their flippers are oriented in different directions, as if posing for the camera, as indeed they are. William Saville-Kent published the picture in his book on Western Australia, *The Naturalist in Australia* (1897). He had the idea of decorating the dining table with newly hatched turtles, but it is more than mere decoration. He wanted to demonstrate the potentialities of the camera "for rendering marine zoological subjects in a state of active locomotion in their parent element."[5] The turtle reefscape would combine *"utile cum dolce"*—as the Latin poet Horace said in quite another context—useful knowledge with beauty, in the dinner setting. I was reminded of the marine turtle exhibit at the Great Barrier Reef Aquarium. Each year aquarium staff raise a few hatchlings from eggs, feed them well on a diet of seaweed jelly and minced prawns, and keep them as exhibits until they are about a year old, when they are returned to the sea. The exhibit performs a similar function to Saville-Kent's soup tureen, allowing viewers to see the characteristic attitudes of the flippers as the hatchling swims.

Raine Island is a coral island littered with the sun-bleached bones of dead turtles. The island is located at the eastern entrance to Torres Strait on the extreme edge of the reef, one of the last of the Great Barrier Reef islands and the first of the oceanic islands of the Pacific Ocean. It is the most important breeding island in the world for the green turtle. When turtles come ashore to lay their eggs, in the laying season between October and December each year, Raine Island is transformed. One estimate has it that on one famous night in 1974, 11,800 turtles beached themselves on its shores.[6] Nowadays people are not allowed to go onshore unless they have good scientific reasons and a permit. Even then, only five people are allowed onshore at any one time.

In 1992 I was fortunate enough to visit Raine Island with a group of scientists and others from the University of New England. As we anchored late in the afternoon, the sea around the island seethed with the dark shapes of turtles. Turtles stuck their heads up out of the water to breathe. Turtles copulated in turtle fashion, clinging together in the water, rising from time to time to breathe while still at it, reputedly able to continue for days. Turtles circled the island, waiting for dark to come so they could beach themselves in the cool of the evening to lay their eggs. The beach was covered with turtle tracks from the egg-laying activities of previous nights.

Reefscape with turtles—all stages of life and death of turtles: live turtles, dead turtles, turtles in various stages of dying, death and desiccation. Any place where thousands of turtles come up each night will have its casualties. The island has its own turtle death traps, small cliffs down which the turtles tumble and overturn. Some of the cliffs are natural; others are the result of the quarrying for compacted guano stone, which occurred in the 1890s. The adult turtle that overturns can't right itself and dies in the heat of the day. Turtles tumble on top of other turtles. Turtles die on top of dying turtles. They lie in the sun, their internal organs hanging heavily upside down within them, overheating horrendously until they die.

The stages of turtle decomposition are as follows: day one after death, the turtle bloats; day two, the head stiffens and comes up, followed by the flippers on day three; day four, the turtle explodes; and from day five it desiccates. The skin dries and stretches over the bones. Later the skin falls off, the plates and bones are exposed and bleach in the sun. Eventually, after about 10 years, the skeleton completely weathers away. Raine Island has turtles in every stage of decomposition and desiccation. Geoff Miller, a Queensland National Parks researcher, was on the island when we visited, and he told us all about it. Miller's research on turtle fertility involved turning the living turtle over and examining the ovaries by means of an endoscope. From the scars he can tell how many laying seasons the turtle has been through. If no scarring, the turtle is laying for the first time. The numbers can be compared, to get an estimate of the reproductive health of the population. If there is an increase in the number of females laying for the first time, this indicates a decline in the number of older females. They may be being taken for food in the islands to the north—and in an unsustainable fashion.

People allowed on the Raine Island shore that day reckoned that they saw 500 dead or dying turtles and that was without seeing what was happening in the center of the island. It is totally forbidden to visit the central depression, which was artificially created by the removal of tens of thousands of tons of guano back in 1890–1892, because it is an important breeding ground for brown and masked boobies. The destruction of wildlife during the mining years must have been terrible, but it has left a flat, open space that some birds now prefer. The ornithologist Brian King reckons that over 6,000 pairs of brown boobies nest at Raine Island.[7] At various times of the year other bird species include the red-tailed tropic bird, the rufous night heron, and frigate birds. The food of choice of the rufous night heron is newly hatched turtles. Twelve weeks after they are laid, the hatchling turtles emerge from their shells and make for the sea. The rufous night herons eat well.

Rock-pools close to the shoreline provide further hazards for turtles returning to the sea after laying. If they get trapped in these pools as the tide goes out, they can be dangerously overheated. It took four men from our shore party to lift some of the trapped turtles out into the open water. They could do nothing for the dying turtles farther in on the island except feel pity for their slow deaths.

Raine Island is notable for the ruins of a 30-foot-high, round white tower, the earliest surviving stone structure built in Queensland. It was built in 1844 by convict labor brought by the HMS *Fly* because of concern about the number of shipwrecks in the region. It was intended as a beacon to alert mariners to the entrance to Torres Strait and as a safe passage through the reef as well as to provide a place to store and gather water. The timbers and tank for the structure were salvaged from a wrecked ship, the *Martha Ridgway*. The round tower was constructed from blocks of coral rock cut from the island. It was visible from the mastheads of ships some 12 miles away. The dome of the tower has now fallen, and the tower has the stench of dead and roosting sea birds about it. Visitors have carved their names into its walls.

Today only researchers are able to visit Raine Island. But others interested in the turtles in this particular reefscape can see them in *Blue Wilderness*, a film produced by underwater photographer-naturalists Ron and Valerie Taylor.[8] The two filmmakers visited Raine Island and were appalled by what they saw of the dead and dying turtles. Valerie Taylor posed the question: "How far should we go in helping these animals survive?" In her reply she made plain that: "We should do everything we can to set them on their path." She wants to see work carried out on the island to remove the cliffs that trap the turtles and to restore the island to its natural state. At least rescue teams should be mounted to save the turtles. The images in the Taylor film are stark, and the conservation message is strong. Ironically, the film crew itself had not obtained permission to land on the island. As Geoff Miller explained to them (and they filmed him too): "It's a place for animals, not for people. By legislation, by habitat."

Around the world the greatest threat to the green turtle is people, say the Taylors: 100,000 turtles are killed each year for food or shell in northern Australia, the Pacific Islands, Papua New Guinea, and Indonesia. Geoff Miller argues that the turtle deaths are part of the process of natural selection: "Their genes are lost. It's nature's way." But Valerie Taylor was unconvinced; she wanted to save those particular turtles at Raine Island, there and then, when they needed it.

The nature documentary has changed between the Monkmans' and the Taylors' time. Both films show dead turtles, but each reflects the attitudes of their times to the deaths. When the Monkmans made their movies of reef life in the 1930s, they camped on a coral island for months and lived off the reef, eating turtles and turtle and sea bird eggs. As late as 1960, turtles were harvested, legally, for food at Raine Island. Only as recently as 1981 did the island become the sanctuary it is today.

The turtle bears the burden of its uniqueness. On the family tree of evolution, it dates back to the beginnings of reptile evolution in the Carboniferous era. Only recently have people learned to find value not only for the turtle in itself but also in its history as a unique living representative of an ancient independent lineage.

Raine Island is a place where the Darwinian struggle for existence is relentlessly played out, with a little interference in the past from humans. I'm not so sure I'd go along with Geoff Miller's story of natural selection happening there, in the sense of turtles having "good genes" or "bad genes." It isn't so much what is "in the genes" that determines which animals live and which die in this place. In the struggle for existence, as Charles Darwin said, death falls disproportionately on the very young. When the turtles hatch, the sea birds dart in and ensure their own survival at the expense of another's young. Only two in 100 hatchlings survive. Of those that do, one in 10 will survive to maturity. Survival in the struggle for existence involves good luck in escaping accidental death. On Raine Island it is often fatal for an adult turtle to choose this path and not another if it falls down a cliff onto its back. The hatchlings are small enough to right themselves if they flop over, but the adults cannot.

Survival in the struggle for existence also means winning the competition for scarce resources: here the competition is for a patch of sand to lay eggs and for the eggs to stay covered until they hatch. What one turtle has so laboriously laid and covered, another may dig up when she begins to make her hole. Charles Darwin would have had some explaining to do, to work out why, in the struggle for existence, so many—obviously *too* many—turtles come back to the same place to breed time and time again. It would be to their advantage to choose a less overpopulated place, with a less hazardous terrain, at least from our point of view.

What this heaving landscape of dead and dying turtles brings to mind is mortality in general, the transience of life as mirrored in the turtle's. There is a solitary grave on the island, the grave of Annie Eliza Ellis, who died there in 1891 at age 52. She must have lived a hard life on the island, as the wife of the guano-mining overseer. Set in a landscape of turtle bones, her gravestone bears a salutary message: "Reader! Be ye also ready." But the gravestone also provides shade for the survival of the chick of a pair of brown boobies that nest beside it each year.

A turtle skeleton wedged under a bush provides a hiding place for hermit crabs during the day. They pile one on top of the other under the jumbled bones, seeking to escape the heat. At sunset the white plates of the skeleton glitter with orange as the crabs stir within and emerge from all orifices. Their bright orange-red claws bear them along the beach. Borrowed white shells protect their soft backs. When one shell is outgrown, it is abandoned and another found to take its place. In the midst of death at Raine Island, we are most definitely still in life.

# Sea Grass Harvest

I was the only passenger that day on the catamaran from Cardwell on the Queensland coast to Hinchinbrook Island. The storm clouds were low and threatening. It had been raining for four months, the skipper told me. But, for the good news, he'd spotted some dugongs on an earlier trip near the mangroves of Missionary Bay. "That's what you look for," he said, handing me a laminated poster of a dugong at the surface of the water. Most of its long plump body was under the water, with just the nostrils poking above. In the illustration the water was clear, so each feature of the dugong was easily visible. Some contrast indeed with the vista before me: the water in Hinchinbrook Channel was murky with sediment after months of rain. As I looked through sheeting rain to where the ghostly outlines of steep mountains met the ocean, I knew I'd be lucky indeed to catch sight of anything in the water.

Dugongs belong to the scientific order Sirenia, so named from the sirens of Greek legend. The dugong is a solid animal—gray, plump, a little hairy—one of nature's more roly-poly creations. The body tapers to a flat tail, fluked, or split in two, at the end. The dugong must rise to the surface to breathe at least once every eight minutes. It does so cautiously, with only the nostrils appearing above the water. When it dives to feed on sea grasses, the nostrils are covered tightly with valvelike lids. The eyes are on either side of the head, as are the ears. The adult is up to 10 feet long and weighs up to 900 pounds.

Dugongs have messy eating habits. As they feed, they often rip up the sea grasses, roots, rhizomes (horizontal stems), and crustacean inhabitants all. They leave a cloudy, muddy trail that can be seen on the surface of the water. An adult dugong eats about 62 pounds of sea grasses a day, while an adult turtle eats about 4 pounds a day.

On my trip across Hinchinbrook Channel it was too murky even to see the mud trails from feeding dugongs. The dugong poster was wildlife promotion hype; I wasn't going to see anything. I said as much and the skipper agreed. "They're timid, shy animals. They don't breach the water like whales or dolphins, but they're there. If you're lucky, you might see a fluked tail disappearing as they turn and dive." That day I had to settle for the next best thing, the frisson of delight knowing that dugongs swim invisibly all round. On one reef brochure the dugong is in fact one photographed in Vanuatu, a dugong of legendary tameness. The story goes that the Vanuatu dugong is a male that has lost its mate and so it approaches tourists in the quest for love.

Sea grasses are flowering plants that have evolved from the land to live in the ocean. The shallow tropical coastal waters of the Great Barrier Reef Marine Park shelter some 2,000 square miles of sea grasses between the coast and the reef, where they act as a buffer zone between the two systems. Rivers run into the sea and bear sediments and nutrients, which might harm the reef. But the roots and rhizomes of the sea grasses trap the sediments and stabilize them. Sea grasses are like land grasses, absorbing nutrients from the mud; in summer or autumn they flower and broadcast their pollen to the sea; they produce oxygen; they shelter juvenile fish and prawns, crabs and worms, in a grass-scape of tiny marvels. The sea grass beds are vitally important to the fisheries, as the sole nurseries for juvenile brown tiger prawns. Here, where the forces of the oceans are tamed by the sheltering coral reefs, the mangroves and sea grass beds develop and stabilize the shoreline.[1]

When I started researching reefscape, I began with thoughts of corals and fish. Only later did I come to appreciate the reef from the point of view of sea grass communities. For sea grasses in shallow coastal waters, the Great Barrier Reef is like the Great Wall of China, holding back the forces of the world outside while maintaining calm for the benefit of the inhabitants within. I'd love to see a film of the life forms hidden in these underwater meadows, perhaps a Great Barrier Reef equivalent of the French documentary *Microcosmos*, which took viewers on a voyage among the microscopic creatures that lie hidden in terrestrial grasslands. I'd not require the film to be made in Hinchinbrook Channel, though, as it is too hazardous for diving there. The water is too muddy for filming, and, worse, crocodiles lurk in the mangroves of Missionary Bay and the nearby Cardwell River. Abandoned fishing nets present other dangers, both to divers and sea life. "Ghost fishing" is the term used for the lost gear that continues to capture and kill fish. Abandoned nets are one reason the diver straps on a knife, to use when entangled.

Because prawns and fish congregate in sea grasses, these places are heavily fished. The 1990s saw tensions escalate worldwide about unsustainable practices in the prawn-trawling fisheries. In trawling for prawns, heavy trawl gear is dragged across the seabed. If an area is repeatedly trawled, the equipment causes great damage to life both on the seabed and above. There is a serious "bycatch" problem if, as may be the case, prawns make up only about 10 percent of the catch. This means that, if the Queensland prawn catch is 7,000 tons annually, the other 90 percent of the catch in the trawl nets—some 63,000 tons of sea life—is discarded and returned to the sea, usually dead or dying. The bycatch might include endangered species such as green and loggerhead turtles, but juvenile fish and sea snakes may also be destroyed. According to 1996 figures from the Queensland Fisheries Management Authority, some 5,000 turtles are caught in trawlers annually, of which some 50 to 70 die.[2] Turtles, rather than dugongs, are the victims, as dugongs seem to keep clear of the noisy trawlers. Each trawl across the seabed also removes 5 to 20 percent of the sponges, sea whips, and sea fans that grow there as well as the smaller creatures that live on these animals.[3]

In 1999, concerned about the unsustainable nature of trawling practices, the Australian federal government legislated for the use of trawl nets with compulsory bycatch reduction devices and turtle exclusion devices, technologies that permit nontarget species to escape if caught up in a trawl. Use of these devices gives a 15 to 90 percent reduction in bycatch. Illegal trawling in the marine park is a considerable problem, soon to be tackled with greater use of remote monitoring systems, with position-fixing transponders fitted to every trawler.[4] Recreational fishers know the importance of the sea grass meadows as fish nurseries and are lobbying hard for a reduction in the damage caused by bad trawling practices.[5] Coastal communities are also well aware of the importance of the sea grass meadows in their areas. People living at Hervey Bay were greatly concerned after severe floods in 1992 killed off nearly 400 square miles of sea grasses, either through the action of storm waves or from light deprivation when muddy floodwaters swept out into the bay. Local residents reported dugong deaths and migration and a decline in the commercial fish catch of 40 percent.[6] As a result of this concern, sea grasses now have their own support group in *Seagrass-Watch*, a cooperative effort between locals and scientists in the Whitsunday and Hervey Bay regions of the reef. Together they monitor the health of sea grasses by regular sampling, so that any changes may be swiftly spotted.

As the lives of sea creatures become better known through filmed wildlife encounters, some of the larger ocean inhabitants are coming to enjoy the status of "charismatic megafauna"—whales, dolphins, turtles, manta rays, and the dugong. "Charisma" is a word derived from Greek, meaning "gift of grace" in the spiritual sense. The word was introduced into sociology early in the twentieth century by Max Weber to explain the attraction of political and religious leaders. People who follow these leaders confer on them charismatic qualities that inspire devotion to their cause, and this perception of charisma may then contribute to great social changes.[7] In the biological world, charisma also counts, and this is something only recently recognized in zoology. In classifications, zoologists normally seek to describe animals objectively. Thinking about what animals *mean* to humans is a new trend in science, though long a feature of myths, cosmologies, and indeed of the value people place on companion animals in everyday life.

Charisma is a quality in the human-animal relationship. As in all relationships, the allocation of grace is haphazard and arbitrary. Being large, scary, and black and white apparently helps, as evidenced by our special regard for zebras, giant pandas, whales, and penguins. Even the black-and-white great white shark could be on its way from hated and feared monster to charismatic megafauna because knowledge about shark behavior is now feeding into public perception in a positive way. The crown-of-thorns starfish is a good example of *negative* charisma. A large creature that mobilizes itself in plague numbers to eat reefs voraciously, it skulks in the mythic dimension of awfulness, as once did the great white shark.

The whale is a recent arrival in the charismatic list. Fisheries expert Daniel Pauly is delighted with the human construction of the whale's charisma even as he is critical of it. He lists the elements in the new mythology concerning whales: whales are beautiful, more so than most other animals; whales do not grunt, they sing; whales have big brains, so they are intelligent; whales want to be left alone to celebrate nature. It is this new mythology, Pauly says, rather than strict scientific evidence, that has created the political pressure in some Western countries, including Australia, to ban whaling. Fisheries science has long argued that whalers were killing too many whales and that their numbers were dwindling alarmingly. But the whaling industry, says Pauly, consistently refused to accept cautions from science.[8] Reasoned argument in the cause of conservation was not enough. It was the whale's newfound charisma that forced governments to act. People today are telling each other stories about encounters with whales, stories that go beyond the scientific into the mythic realm. A similar story may be told for attitudes and beliefs about the dugong.

Before the dugong had charisma in European eyes it was called a "sea cow," in comparison to the familiar domestic animal. "Sea cow" has connotations of a no-nonsense utilitarian view of the creature. When dugongs seemed plentiful, and threats to their continued existence were unknown, a small dugong oil industry existed, briefly, on the Queensland coast in the 1950s. At this time a dugong fisherman named Snow Goring, from Gladstone, employed Indigenous hunters to kill the animals, which he boiled up for their oil. His explanation for the term "sea cow" is: "You butcher them like a cow, straight up the center."[9] Dugong oil extracted from the carcasses was believed to have good healing properties and a penetrative power great enough to leak through the glass of bottles. It was used in cooking, as it did not burn. Snow Goring reports that dugong tastes like pork or beef, an "off-pork, off-beef taste." The fat was very sweet. As a close-fibered meat it made good bacon. One lugger worker favored the pressure cooker to overcome toughness.[10]

I started this story with tales of dugongs and sea grasses, stories spun from science, the history of cooking oil, and belief in charismatic gifts. Other stories come from "the time before this time," when the Hinchinbrook waters were fished differently. On the channel side of Hinchinbrook Island, at Scraggy Point and Missionary Bay, there are large stone fish traps, relics of lengthy occupation by Aboriginal people. These fish traps are an elaborate series of low stone walls built out into the sea. Fish follow the paths they usually take to shore with the rising tide and swim over the walls into the traps. The traps are intricate, with raceways, loops, funnels, breakwaters, and arrowheads to direct and trap the fish. As the tide falls the fish are stranded, either out of the water or in an enclosed pool. The first dated use of the fish trap constructions is 2,000 years ago and they still work today.[11] They are better described as "automatic seafood retrieval systems," says archeologist John Campbell in a phrase that emphasizes their technological complexity. Considerable knowledge of fish behavior, and the flow of the tides, was needed to make them. Not only did they store fish till needed, but also the traps themselves encouraged further growth of food. Shellfish grow on the stones and help cement them together. Edible mangroves take root. At the time of initial contact with Europeans, Hinchinbrook Island supported an entire Aboriginal kin group.[12]

The modern tourist resort on Hinchinbrook Island stresses its offer of "magnificent solitude" as a place of retreat from all the intrusions of work.

I traveled as a tourist across Hinchinbrook Channel. I didn't see a dugong. Instead, I studied another poster on display in the catamaran. It had a definite message: look out for dugong, take care of them because they're vulnerable to extinction. I looked more closely at the poster. It included images of dead or dying dugong. One image showed a dugong lying on a beach, its large fleshy mouth parts pulled back to expose the sea grass grinding disks. Another showed Indigenous dugong hunters in a dinghy, holding the dugong with its head underwater, a rope around its tail. This is the traditional method of killing the dugong, once it is harpooned and pulled to the boat. With its head underwater, it soon drowns.

The text of the poster spelt out a conservation message, while the images showed the story of Indigenous hunting practice. Here was the dilemma: the concern that the dugong is endangered is contrasted with pragmatic realities that until quite recently dugong was taken on a regular basis in Australian waters for meat and oil by Indigenous and non-Indigenous hunters. Then it was assumed the numbers were high and so it seemed to be. Because dugongs live largely hidden lives, only recently have people realized that their numbers are declining drastically along the Queensland coast. They are slow-breeding and long-lived animals, achieving lifespans of 70 years or so. The female may become fertile some time after 10 years of age and will perhaps have one calf only every three to five years. She will carry the calf for some 12 to 13 months, then lactate for 18 months. Even after the calf can forage for itself, it may hang around till the next calf is born. Population decline can happen quite sharply, if breeding-age females are taken.[13]

Keeping track of dugong numbers is difficult work. Scientists at James Cook University have been doing regular aerial surveys of dugong herds since the 1970s. They use a spotter plane flying 450 feet above the water, with two teams double-checking, as they count the dugongs coming to the surface to breathe. Later they have to convert the count to population estimates, using the full complexity of statistical analysis.[14] The technique spots trends rather than absolute numbers. North of Cooktown the populations are thought to be stable. South of Cooktown the trend in some places is 50 percent down on earlier figures. Accepting these figures, the Indigenous Councils of Elders, which control traditional hunting, have suspended hunting south of Cooktown.

Dugongs are also caught and killed in commercial fishing nets. Gillnets are square-meshed nets made from fishing line material that are held vertically in the water by means of a lead line along the bottom of the net. Fish attempting to swim through are caught by the gill-covers. Dugongs get caught by the flippers, nose, or tail in large-meshed nets and soon drown, unable to free themselves from the strong nylon fibers. Today, the danger is recognized, and gillnetting has been restricted in certain dugong protection areas along the Queensland coast.

In 1996 commercial fishing in the Great Barrier Reef Marine Park was worth some $75 million annually, while prawn trawling netted some $78 million.[15] Scientists working on dugongs find their work enmeshed in controversy. The pressure is on them to prove there is a problem, and their results are always contested. Recently dugongs have been fitted with electronic tags and tracked by satellite.[16] Little do these satellite-linked dugongs know that, as they grub about their sea grass rhizomes and roots, they have their aerials in the information age.

A 1998 Australian Broadcasting Corporation *Quantum* television program "The Dugong War," examined the political fallout from the federal government's introduction of the first dugong sanctuaries along the Queensland coast. In these sanctuaries, as initially envisioned, there was to be a complete ban on gillnetting. Following protests from commercial fishing interests, the government modified the term "sanctuary" to a "Zone A/Zone B" proposal in which gillnetting would be allowed in both zones under certain restrictions. The Australian Broadcasting Corporation program told the story of the conflict between fishing interests and scientific research. Time will tell if this legislation will be, as the federal government proudly proclaims, its greatest scientific achievement. Time, however, is what the slow-breeding dugongs may not have, nor do they understand the meaning of zones drawn by legislators on maps. They move in and out of Zone A/Zone B as they please. The scientists who offered their hard-won scientific advice were dismayed by the compromise.

As I watched the television program I was struck by the irony of the dugong chase. The scientific pursuit of the dugong has much in common with traditional Indigenous hunting, though the outcome is different. Men in a boat chase a dugong, following the sea grass trail. To attach the radio-tracking device to the animal, it must be held close to the boat and the dugong must be captured. As they approach the creature, a strong man with a noose lunges forward to snare the dugong around the neck and bring it close in to the boat where the radio transmitter is fitted. As I watched, what came forcefully to my mind was footage of dugong hunts from early documentary films, where killing by Indigenous hunters was staged for the filmmaker. As one old-time north Queensland guide said of a film he helped make in the 1960s: "The Italian film team wanted a dugong so we got a permit and they got a dugong."[17] In another film made in the same period the Indigenous hunter stands in the bow of the canoe, three-pronged spear in hand. He judges the right moment, then throws himself and his spear into the water in spectacular fashion, so that the spear hits the dugong with the additional force of the hunter's body. The spear is attached by rope to the canoe and the dugong is pulled in.

Comparing the attitudes reflected in documentaries of the 1960s and 1990s shows how attitudes toward wildlife conservation have changed in a short space of time. In 1966 scientists from the Scripps Institute of Oceanography in La Jolla, California, came to north Queensland onboard a research vessel, the *Alpha Helix*.[18] In the name of their scientific research, they also filmed a dugong hunt. Two Aboriginal men chased and wounded a dugong with a spear and then towed it to a canvas holding pool on the shore. Blood flowed freely in the water, in Technicolor. In the canvas pool, scientists carried out respiration experiments on the wounded dugong. The scene cut to a scientist who walked out of the water carrying a dead baby dugong in his arms. He placed it on the sand and the scientists bent over it, trying to revive it. They pushed on its back to perform artificial respiration. They recorded this act on film. If they had revived it, what would have been its fate? Perhaps a few more experiments on dugong respiration in the baby and then certain death. The dugong infant needs its mother for some 15 months. Traditionally, hunters preferred taking the female with the calf: the calf was wounded and brought to the canoe, where the female followed in response to its distress calls. Watching the *Alpha Helix* film some 30 years after the events it recorded took place (which also included killing and eating a carpet snake), I wondered what they thought they were actually recording for posterity on the film. The *Alpha Helix* scientists must have thought what they were doing was good and worthy of record. Today it seems pointless and cruel.

Ten thousand years ago people started to cultivate the grasses that grow on land and found increasing uses for their seeds in food. Soon it may happen that underwater farming will come to pass, as researchers gather the seeds of sea grasses and learn more about them. Some sea grasses release their seeds directly into the sediment, where they may remain dormant for years before germinating. This is why some sea grass beds, when damaged, seem to recover as long as enough parent plants remain as protection for them. Sea grasses and crabs have learned to live together in a relationship of mutual convenience: crabs and prawns help bury seeds as they burrow through the sediment.[19] Gardening sea grasses, by adding a dose of garden fertilizer, has even been trialed. Doctoral student Jane Mellors added fertilizer to her sea grass garden plots to study how grasses maintain the quality of water by removing nutrients from it.[20] Who can tell where this research will end—perhaps with underwater paddocks of new cereals for the bread of the future?

Until the end of the twentieth century, the resources of the sea were taken, as once in human history hunter-gatherers foraged in the wilderness for the resources of the land. What if a similar change to agriculture and farming occurs soon with marine species? Aquaculture of reef fish and prawns is being promoted as a logical future progression: just as people once domesticated cattle, sheep, and chickens, so, it is claimed, it is the turn of prawns and reef fish to enter an era of rapid domestication. Aquaculture is still a small industry in Australia: with 60 species of fish, shellfish, and aquatic plants farmed, the industry in 1998 was worth some $220 million annually. Critics are swift to point out, correctly, that many aquaculture projects undertaken internationally have been environmentally disastrous. Peter Rothlisberg, leader of the CSIRO's Aquaculture and Biotechnology Program, acknowledges this: "We've certainly seen lessons from around the world where people have gone in and, say, cleared mangroves to put in ponds for fish or prawns. That's not sustainable practice. It's not good for the coastal zone. Those mangroves and sea grasses support wild fisheries and other important animals, and it's not good habitat to put ponds in either. The soils are unsuitable. The drainage is poor." Australia, he says, will learn from those mistakes and not repeat them. While it has been 10,000 years since the beginnings of terrestrial agriculture and 200 years since the origins of scientific selective breeding, the next stage to working sustainably with marine food sources will be much swifter, Rothlisberg estimates: "We don't have to wait 200 years to get the benefits of this. We can start right now."[21] Soon domesticated species of fish and prawns will look as different from their wild counterparts as domesticated cattle and sheep.

In early frontier days in Australia, explorers went prospecting for gold. Today, marine bioprospectors are searching under the seas for new riches. They are looking for marine invertebrates and plants that will provide the basis for new drugs, particularly anticancer drugs. Many coral reef species have developed ingenious forms of self-protection through the secretion of toxins, and some of these compounds, developed naturally over millions of years, have the potential to kill cancer cells. Dr. Patrick Colin from the Coral Reef Research Foundation in Palau says: "We are mostly interested in soft-bodied sessile invertebrates which rely on their chemistry, rather than stinging cells, spines, jaws or teeth for survival. Such animals usually have compounds that make them unpalatable or toxic to potential predators or that can be used to kill neighboring animals in the competition for space."[22] His work is funded by the U.S. National Cancer Institute. The defenses that marine creatures build against each other may hold the clue to cancer treatments for otherwise intractable solid tumors.

Western medicine is taking a fresh and unprejudiced look at the use of marine animals in traditional medicines, such as Chinese medicine. Of course, there is the terrible threat of extinction of wild species, such as seahorses, if too many are taken. But it is important to find out how traditional seahorse remedies might work in treating ailments such as arteriosclerosis, impotence, and thyroid disorders. Canadian researcher Amanda Vincent says it is wrong to simply reject all traditional medical practices as nonsense. She calls for a better understanding of traditional uses of marine invertebrate animals in order to document their trade and promote better management.[23] Another way to tackle the problem is to work out how to breed seahorses in captivity to supply the traditional medicine trade in a sustainable and cheaper fashion than wild harvest. Or use biotechnology, where, once the effective ingredient is isolated, ways are found to grow it in the laboratory. The union of aquaculture with biotechnology is still in its early stages and will no doubt prove as controversial as biotechnology has already proved itself to be in terrestrial agriculture.

I began this chapter with reflections on dugongs and the sea grass meadows of the Hinchinbrook Channel. I shall now travel north and west to the Yanyuwa people of the southwest Gulf of Carpentaria. Anthropologist John Bradley has researched the cultural and spiritual meanings that the Yanyuwa attach to the dugong and other sea creatures. According to the Yanyuwa people, dugongs, sea turtles, and sea grasses exist in a relationship of mutual benefit to each other and to the people who hunt them. The Yanyuwa people have a word in their language for this relationship, *walya*. This word, says Bradley, has no direct English translation but is generally translated as "dugong and sea turtle." It embraces the notion of these two creatures, not only as the largest marine animals hunted, both of which feed on sea grass meadows, but also their significance in the Yanyuwa spiritual tradition. It acknowledges that dugongs and sea turtles belonged to the ancestors and also to the Yanyuwa people: they are symbols of continuity with the past. *Walya* enfolds the mythic past within the present: it preserves the links people have with their ancestors; it links animal behavior to habitat and hunters.[24]

Yanyuwa people say that just as humans observe and interpret the actions of living things, so living things such as dugongs and turtles observe and interpret the activities of humans. Yanyuwa people say that dugongs and sea turtles have their own law, their own culture, and their own way of being. Dugongs and turtles are seen as kin to their source of food, the sea grass. Dugong hunters are kin to the dugong. The source of this kinship is with the ancestral beings and their creating the country in the mythic past, in "the time before this time, the first time"—which has come to be called the "Dreamtime" or the "Dreaming."[25] The Spirit Ancestors, who include the Dugong Spirit Ancestors, traveled across the country, leaving their mark in landscape features such as rocks and reefs. "Country" here includes "sea country," the land under the sea where the sea grasses grow and which the Dugong Spirit Ancestors passed through as the dugongs do today. The sea itself is a Spirit Ancestor, and waves near the shore are associated with the Spirit Ancestor of the sea snake.

The song of the Spirit Ancestor Dugong Hunters, as Bradley records it, is both a list of the named places through which they passed and an affirmation that their act of creation still resonates in the lives of the people today. The "first time" lives on in the names of places and the maintenance of traditional celebration of these sacred places. By celebrating the mythic past, present-day hunters identify with it, maintain it in the present, and preserve it for the future. To hunt is to bring food to others. Hunting is skillful; it invokes the thrill of the chase; it has overtones of machismo. Hunted animals suffer pain; ways of killing are not swift. The hunter, in watching the animal, knows the relationships of the animal to its habitat, knows the habits of the animal in order to strike it. The hunter is the ecologist.

To know nature in order to hunt, to love nature in order to kill—these are uncomfortable facts. Along the coast of eastern Cape York, as in the Gulf of Carpentaria, there are many sacred sites and story places connected with the dugong and the turtle, sites that similarly link people with animals, plants, landscapes, and seascapes and with events from the Dreamtime. Over the last 70 years Aboriginal people along the Cape York coast have been forcibly relocated to camps away from the sea and their traditional sea estates. Despite this, customary hunting and fishing have continued, and this continuity makes a link with the past that has profound meaning for Aboriginal people.[26]

In the scientific tradition, dugong and whale watchers tell stories about the animals they study. These stories are placed in the context of a theory of evolution and evolutionary advantage of traits such as a natural curiosity. Science relates the evolutionary record of interdependence. Both scientific and Yanyuwa ways of knowing meet in the recognition of the differences between humans and animals, yet also the continuity between them. However, there is one important discontinuity that science presents: the concept of the extinction of species. The history of the evolution of species is also a record of past extinctions. Yet this concept may not exist in other cultures in which that species has always been plentiful, and indeed it is sometimes hard to find evidence that the concept is much understood outside of science itself.[27] Along the Queensland coast, fishermen take fish in gillnets strung across river estuaries in areas adjacent to the Great Barrier Reef Marine Park. As each fisherman goes about this daily legal or illegal business, each would know what happens in their nets. It may be that some fishermen are vigilant and never kill dugong or turtles and thus are righteously indignant at any legislation that threatens their legitimate activities. But destruction happens, and it is the cumulative effect of similar kills up and down the coast that matters. As I watched the Australian television program on "The Dugong War," I had to ask myself whether any of the nonscientists interviewed appreciated the issues. When short-term business gain or making a living is what matters, extinction of a species that might take place over 200 years is outside the sphere of consideration.

Whales are returning to the Great Barrier Reef, but they had the wide ocean and the cold waters to which the survivors could retreat in order to come back. Whales are returning, and many people feel tremendous pleasure at their survival. When whales beach themselves accidentally and are in danger of dying, stranded on land, people arrive in droves to try to help as best they can. Images of large stranded and dying whales provoke emotions of pity for them, where not so long ago whale killers might have swiftly moved in to make money from the carcasses.

Dugongs are not nearly as widely dispersed as whales throughout all the oceans of the world. In the space between the sea grass and the satellite, between Indigenous traditional ways of life and Western consumer societies, lies the path of "progress" and more of its discontents. Progress in fishing techniques and modern farming practices, the growing popularity of boat ownership, and increasing development along the Queensland coast all adversely impact the sea grass meadows and the dugong's survival. Better for the dugong were it able to tolerate colder water, like whales, or had a more ocean-going inclination. There is nowhere else for them to hide except the shallow sea grass meadows that they are now compelled to share with fishermen and tourists.

If it is considered macho to hunt tigers, bulldoze forests, and kill 40,000 whales a season, conservation measures won't work. Nor will they work if greed reigns supreme.[28] If room is allowed for mythologies to flourish, whether these are new or ancient versions, there is some hope for conserving large marine animals. The fact is that for a variety of reasons, some irrational, charismatic animals are evoking an immediate personal response from increasing numbers of people: whales sing; turtles shed tears; dolphins have powers of healing; dugongs vibrate in sympathy with celestial harmonies. Each of these statements may be both true and not true. Mythologies matter in the charisma game much more than facts. Science itself will benefit from new mythologies because the objective detachment from organisms, habitats, and systems that scientists study is itself another myth.[29] The very concept of "charismatic megafauna" is a step in the right direction: it recognizes that perceptions matter in zoology as in life.

# Chick City

The Chesterfield Islands lie in the middle of the Coral Sea, nearly halfway between Noumea and the Great Barrier Reef. Once British territory, the islands were named in 1793 by the captain of the British whaler, the *Chesterfield*, but since 1877 they have belonged to France. They are so remote from both Noumea and Australia that they have been relatively undisturbed, apart from the casual destruction of wildlife by passing mariners, and some guano-mining activities in the nineteenth century. For most of the twentieth century the islands have been left to the birds. The scars of the guano mining have faded to leave an almost natural reefscape. Once I was lucky enough to be there: December 1995.

The Chesterfield Islands are coral islands that have grown to the surface of the sea from the rim of a subsiding sea mountain. Two rings of islands and reefs lie adjacent to each other, with reefs and coral islands at all stages of growth, from vegetated islands to sand banks to reefs still submerged and growing toward the light. At Loop Island we anchor in 33 feet of water so clear that we see *bêche de mer* grazing on the sand beneath the hull and the trails of sand they leave behind. At sunrise when the red sky is reflected in calm water and the line of the horizon disappears, I have a still, calm sense of blessedness. Celtic legends tell of the misty westward isles, the place of repose to which the soul is borne after death. My place of peace is an island that shimmers in heat haze, set in a blue-green coralline sea. Here at the Chesterfields we are differently blessed.

As the dinghy approaches the shores of North Avon Island, flocks of sooty and noddy terns swirl up and wheel above the dinghy, squawking loudly at our intrusion. The common noddy has a steel-gray cap on its head, eyes half-rimmed with white, its body a sleek soft black. The white-capped noddy nests in trees. The sooty tern is whiter than black—white below and soot-black on top. Its eyes are elegantly rimmed with black, as if a swathe of soft kohl has been etched across each eye from beak to crown. High in the sky frigate birds circle in the thermals, black wings spread wide, waiting and watching. They are the "middle men" in this fishing business, relying on other birds to do some of their fishing for them. The frigate bird waits on high and swoops when it spots a booby bird returning from sea with fish in its crop. The booby is harried until it gets flustered and vomits up its catch. The frigate bird fishes the easy way, if you like your fish predigested.

The unmistakable whiff of coral cay comes out to greet us, that heady mix of bird droppings and salt air I've come to know. On the high-water line on the beach, near where we land, is a makeshift nest of a masked booby and its equally large, fluffy white chick. They will not move far just because humans have arrived. But if we go too close they will squawk extremely loudly. Father has a high-pitched squawk; mother's is lower and noisier. Baby squawks longest and loudest of all because its cries may mean its survival. Masked boobies might lay two eggs, but if two chicks do hatch, only one will survive. The eggs hatch four to five days apart. It used to be assumed that the older chick survived because it was more successful in attracting its parents' attention for food. After spending long periods in quiet observation of the tough facts of booby life, marine biologist Myriam Preker found otherwise: the older chick will attack the younger one and eject it from the nest, where it survives only two to five days after hatching. The parents seem to bring enough food for two chicks, at least in the early stages of life. It is the behavioral adaptation, the "siblicide," of the five-day-old chick that brings death to the younger and weaker.[1]

Farther down the beach a masked booby chick is feeding. It thrusts its small beak well down into its parent's throat and slurps up liquefied fish. There are two pairs of masked boobies with young close together on this part of the beach. From time to time the adults jab at each other. They have the whole beach on which they can spread out, but clearly to both sets of parents their particular spot on the beach is the best, to be defended against all comers. The chicks vibrate their throats, ululating without noise—just the shiver and shake of down on their long necks. Their mouths hang open. Endurance is all.

Fledged booby chicks sit on the water and dip their heads under, peering around as if checking if their feet are still there. A juvenile frigate chick, brown-faced, hook-beaked, perches on an argusia bush, waiting to be fed. It is already a strong flier but will not achieve full maturity until it is two years old.

I turn my back on the beach, with its groups of bobbing boobies, to face the interior. Here the island dips into a broad, shallow depression where guano was once mined. Sooty terns occupy the entire space, about 15 acres in all. At three nests to the square yard, there must be between 400,000 and 600,000 birds resident here, living in grand hullabaloo. Sooty terns scoop their nests in the sand and prefer sites that are open and exposed like old guano mining sites. Humans created this barren interior. Now it is the birds' turn to create a landscape that shivers with their vigorous motion, that vibrates with their continual noise. The birds feed off the richness of reef waters. Here reefscape meets landscape, the one creating the other.

Another reason for the huge tern population is that there are no seagulls on these remote islands. Gulls eat the eggs or take the young chicks. Gulls and terns belong to the same zoological family, the *Laridae*, a group of large marine birds found in the warm waters of the world. However, there are no bonds of family feeling among their membership. To gulls, just about anything is food, including their young relations.

The tern nests are a casual jumble of sticks, guano, and dead grass. The colony is divided into zones according to the age of the chicks. To one side I see a zone of nests where the egg has not yet hatched and a parent sits on guard. To the other side, on a neighboring patch of dry brown grass, is a nursery zone with one-day-old chicks. I know they are only one day old: the night before they were still in their eggs. The parent birds settle in successive waves of nesting, where first arrivals set up house in one zone. Later arrivals cluster near them. The eggs hatch in waves across the colony.

The one-day-old chicks are wet, soft, gray-and-brown balls of fluff as they unfold from the egg. They rest after breaking through the shell and set their small wings out to dry. Stretching, staggering to rise on thin wobbling legs, they face their new life. They wriggle their claws in the dirt, fluff out their feathers, and preen themselves. They may be new to the game of avian life, but they are born "semiprecocial," that is, relatively mature. The chick hatches with a good covering of down and is soon able to stand and beg for food, mouth gaping. It does not peck the ground, like domestic chickens. Its food comes from above, from the beak of a parent bird.

Chick city.

Chicks stay together at each stage of their development, remaining in the early weeks near their nest site. A parent stays with the chick for about a week, and then it is left for the day with other chicks the same age while the parents return to fishing at sea. The six-week-old chicks flap their wings and attempt to fly in their section of the nursery. Still older chicks leave the nest area and flock near the water's edge. They run up and down the beach in ripples of activity. They look to the sky where at dawn the parents wheeled out in dark clouds of birds, to fish all day at sea. When they return to the island at dusk, there is a flurry of greeting. "Look-at-me, look-at-me, wide awake, wide awake." Parents call to chicks, chicks call to parents. I am stunned by the noise and the flurry of activity and the casual cruelty of the adult birds. A sooty tern in full "colony site defense mode" is a fearful sight. Aggressively, it thrusts its wings back and its head forward, beak jabbing. Sooty tern territory near the nest is strictly demarcated. If a one-day-old chick totters out of its territory, a neighboring parent will attack it. The target of the beak may well be the eye of a hapless neighboring chick. Dead one-day-old chicks are tossed from the nest like rubbish.

The interior of Loop Island is a seething mass of black and white. Sooty terns may prefer the open spaces, but the noddies, both common and white-capped, prefer a little more shelter. Common noddies nest around the margins of the sooty tern colony, under a low bush, or beside the concrete plinth that proclaims the land belongs to France. The white-capped noddy, differentiated by its sleek silvery-white capped head, nests in trees where there is more room to move without harassment. The small straggly casuarina bends low under its weight. The bird wakes in the heat of the day and leans forward to prop its beak on a branch, drooping. A large tick hangs from its face, with smaller black parasites clustered near. Surviving means enduring.

In 1992 a cyclone centered on the islands, and birds too young to fly before the storm would have perished in large numbers. That may have been why we found such a large number of birds when we visited three years later. It was a population of birds in the process of recovery from natural disaster, the sooty terns congregating in all stages of their breeding cycles, from courting couples to harassed parents of hungry adolescent fledglings. In the population cycle of "boom or bust" the sooty terns were enjoying the boom.

If we were to come back to Loop Island later in the year, its bird population will have dramatically declined. The sooty terns stay on the islands until the breeding cycle is complete and then disperse out to sea where they spend most of their time in the air. Their plumage is not water repellent, and if they land on the surface of the sea, it is only for a few brief moments. They skim the surface of the sea and snatch small fish and mollusks from near the surface. Noddy terns, in contrast, have plumage that repels water better, and they are often seen flopping onto the sea, able to rest comfortably far from land.

A broken wing of a marine bird will not set, as the bone is too porous. This quality of the bones, its porosity, which is of benefit to the bird's long periods aloft, counts against the bird when it breaks its wing. The bone will never heal. The bird with the broken wing flaps around until it dies. There are surprisingly few dead birds lying around. Feathers fly off in the wind. The sand drifts and covers the bodies of dead birds. Scavenging crabs clean the bones, as do the smaller invertebrates such as the tiny springtails that live in the sand.

A day or so later at Bennet Cay, I sit on a patch of dried grass near the shore—the grass dead yet alive with small crickets and lice and fleas from the birds. It is getting on for evening, and in the cool of the day two experienced bird handlers move between two species of resident boobies, masked and brown, collecting samples of blood. The chicks of both species are the size of a domestic fowl and are easy to catch. The bird handlers wear raincoats to guard against splatter from bird droppings and regurgitated fish, as one holds the bird and the other collects two drops of blood from under the wings. DNA in the blood will provide information about the genetic differences between the two species of birds. The Chesterfield Islands regularly have three species of boobies that nest there, including the less numerous red-footed booby that nests in trees.

Out near the horizon white waters break on the encircling reef. The sea within the lagoon is mirror flat, the water clear, the sky fading from brilliant blue to pink-streaked gray with the approach of evening. A long sand spit runs out from the end of the cay to a series of small cays that lie beyond, strung out like exclamation points.

We are at the still small center of a universe of sea and sky and birds and yet more birds. Nothing seems more important here than the reproductive strategies of the birds. The rest of the world goes one way, but the birds keep reproducing here as they have always done, ever since these islands first grew from their base in the sea. Where did they nest before the islands came? They may have come to the Chesterfields to get away from humans—and here we are.

The birds live off the reef and in turn help create it. Their droppings are rich in phosphorus and nitrogen, which they deposit onto the coral island rubble. In time the combination solidifies to a phosphatic rock that, in crumbling, provides soil in which vegetation may take root. The birds bring in seeds, in their plumage and their digestive tracts, and these seeds will green the coral cay.

Loop Island has signs of human presence: a weather station is hidden away behind trees and there is the plaque to commemorate, in 1977, the centenary of the *rattachement* of the Chesterfield Islands to France. The French word *rattachement* means "link" and also signifies a relation of hierarchy—as the Chesterfields established *under* France. Even over a colony of birds, France likes to maintain her cultural dominance.

The question the reef traveler of today asks herself is whether she has any right to be in this wild place at all, lest she contribute to the vanishing of all that brings joy here. Nesting birds mark out their territories in ways known only to them. Sea birds harass each other enough when nesting. People disturb sea birds, however careful they imagine they might be, and many will not care to be careful. The population along the Queensland coast is increasing with increasing coastal development, and boats are becoming technologically more sophisticated and faster. Many once-remote islands in the Barrier Reef are becoming more accessible to recreational boating. Recreational activities disturb birds. Soon three-quarters of sea bird islands on the Great Barrier Reef will be within a day-trip range of the coast. There will inevitably be a larger human presence in fragile places. People are always intruders, no matter how they come or why.

In the early days of guano mining on the coral islands of the reef, from Lady Elliott Island to Raine Island, the vegetation was stripped back to get at the rock beneath, and this casual destruction would have devastated the wildlife. Less obvious effects of human intrusion are harder to chart. Sea bird nesting populations will vary naturally from year to year, and only long-term trends will definitely mark a decline. Success in raising young depends on factors such as the availability of food, and this also will vary.

On Rocky Islets one recent set of experiments to investigate tourist disturbance on populations of nesting bridled terns produced ambivalent results. The design problems for the experimenters were: first, how to simulate disturbance by tourists and, second, how to measure its effects.[2] Experimenter Emma Gyuris explained how she and her team walked among the birds pretending to be tourists—talking, laughing, and walking, carefully, not over the birds but close to them. They then measured the hatching success of the birds and the growth rate of the chicks. Yet they found no detectable differences between the disturbed group and a control group of undisturbed birds. In fact the disturbed group had a slightly higher success rate. You'd think it would stand to reason that disturbed birds would suffer. Was there something wrong in the design of the experiment? Was it the case that the birds weren't fooled? They knew these were scientists, acting, pretending to be ordinary tourists? Perhaps the scientists were not violent enough with the bird sites? Perhaps they should have brought their dogs with them and let them loose for a run, as some irresponsible people do.

At Michaelmas Cay, a heavily touristed site, there has been a decline in two of the four main breeding species, the sooty terns and the common noddy terns, so there, it seems, conservationists are right to protest that fragile places should be protected from human intrusion.[3] At the Swain Reefs, where there has been, so far, little human disturbance, significant declines in the numbers of adults and nests of the brown booby are thought to be related to a reduction in the food supply.[4] It seems that the impact of visitors is but one factor and that what also needs to be understood are the population trends of individual species.

Should I have gone onshore on the Chesterfield Islands? Did I create disturbance? I know I came back from this place with a sense of both the fragility of life there and its robustness. The Chesterfields have recovered from the man-made disaster of guano mining as well as the natural disaster of cyclones. These reefs are very remote, so they do not suffer the destruction of inshore reefs with the runoff of fertilizer nutrients from the Queensland coast. When mariners came to islands like this in the old days, as when HMS *Bramble* called at Raine Island in 1843, their first act on landing was to break every egg on the island, so that they could know that the eggs they found next would be fresh. Then they collected the eggs they didn't eat and stored them in casks, and they killed birds for both food and sport.[5]

Once reefscapes with birds meant fresh food; now they mean photo opportunities for tourists—many more tourists than there were mariners.

Reefscape with birds—the impression is one of layer upon layer of birds, each with their preferred level in air from which to plummet into the sea below. Boobies and terns fly in the middle layers, with frigate birds remote in the high layer of air, ready to swoop onto other birds. Boobies are so clumsy on land, so awkward in their takeoff from land to air. But once aloft they fly with minimal wing movement as they take advantage of, here, the uplift of air; there, its fall. When they dive to fish, with wings folded back, they make the dead-weight drop from air, knowing exactly where the fish will be in the water after it has moved on from first being sighted. Then the move is made from water to air again, with the transition like a continuation of underwater flight.

At the Chesterfields another marvel is the fecundity of this reef, how 400,000 birds can find enough fish to live on, as obviously they do. My experience of the Chesterfields showed me there is still some wilderness out there in the world. Wilderness is often described in Northern Hemisphere terms that are irrelevant to those of us in the Southern Hemisphere, such as the distance that can be walked in a forest without seeing evidence of human artifice. Tasmanian Green Senator Bob Brown's definition is "a region of original Earth where one stands with the senses entirely steeped in Nature and free from the distractions of modern technology," and I think this would fit reefscape better.[6] Except, of course, it took modern technology in the form of an ocean-going vessel to bring us to that wild place in the Chesterfields and refrigeration technology to maintain us at a level of personal comfort so that we did not need to kill large numbers of birds for food. We are remote from other lands and people, but still the predominant sense is of the overwhelming presence of the birds. Experiencing this wilderness brings respect for their continuing presence and their capacity for survival.

# An Island in Time

I stand in the rock art site known as the "Ship" rock shelter. I am on Stanley Island in the Flinders group of islands in Princess Charlotte Bay north of Cooktown. It wasn't easy to get here. Tourists must come a fair distance by sea, and this they have in common with other seafaring visitors of the past few thousand years. The art covering the walls of the Ship shelter shows ships of all nations, painted in red and white ochre on the red sandstone: sailing ships rigged in the distinctive styles of the European lugger and the Macassan (Indonesian) prau; a dugout canoe with a figure standing upright in it, hands outstretched. In shelters on nearby Castle Peak there are similar paintings: of a steam ship, and a detailed image of a lugger, identifiable as the *Mildred*, towing a dinghy.[1]

This rock shelter at Stanley Island seems positively cosmopolitan. I turn to look out toward where the ocean lies, now hidden from view by the fore-dune of the beach ridge. If I disregard the boardwalk under my feet, I can easily imagine myself back in the time of the former artist inhabitants, warily on the lookout for strangers from the sea. Travelers might come from the islands to the north, from the places now known as Torres Strait and Papua

New Guinea. There is a long history of contact between these groups, whether in trade or conflict.

The Aboriginal people who once occupied the Stanley Island rock shelters traveled round the islands in single and double outrigger canoes. It is about 5 miles from Bathurst Head on the mainland to Flinders Island, then 500 yards across Owen Channel to Stanley Island. The canoes were poled in shallow waters and paddled across the channels.[2] The watercraft often had only an inch or so of draught between the safety of the canoe and the dangers of the sea. People traveled widely, well outside their country, north to Torres Strait and south along the reefs and islands to where Cooktown lies today. When English naturalist Joseph Beete Jukes sailed along the coast of Cape York in 1843, on the survey ship HMS *Fly*, he reported an offshore encounter with two men fishing from their canoe. What impressed him were the tidiness and organization of their canoe, as if the men he met were like sailors on the HMS *Fly*, keeping it shipshape. The canoe was over 20 feet long, he noted, and made from a hollowed-out tree with an outrigger on both sides. Inside it were stowed all the implements for fishing—a neat coil of rope, spare paddles and outriggers, spears, throwing sticks, twine, fishing gear, large shells for bailing, with "everything neatly fastened into its place by a bit of line."[3]

The Macassans have been coming to northern Australian waters for 400 years or so, when seasonal winds were favorable, to collect what they called *trepang* or sea cucumbers. In 1821 Captain Philip P. King visited Stanley Island as he sailed north, charting the coasts for the British Navy in the interests of colonial power. King gave the islands their European names. Another survey ship, HMS *Bramble*, waited in Princess Charlotte Bay for 10 days in August 1838 for explorer Edmund Kennedy, but he never arrived. Later it was learned that Aborigines speared him near the end of his attempt to lead the first European expedition to travel overland to Cape York from Rockingham Bay near Townsville. The difficult terrain, the near impenetrable bush, the continual rain, and the hostile Aborigines created horror conditions. Only three expeditioners survived to tell the tale.

Onboard the HMS *Rattlesnake*, the companion ship to the HMS *Bramble*, young Thomas Henry Huxley was disappointed at being unable to join the Kennedy Expedition. Huxley was ship's physician, and he was needed onboard. Huxley spent much of his time on the reef shut up in his cabin, in a state of lovelorn melancholy. In love with Henrietta Heathorn, whom he met earlier in the year in Sydney, Huxley wrote next to nothing about the reef. Despite this unpromising start to his career as one of Britain's first biologists, Huxley returned to Britain and became a good friend of Charles Darwin, as well as a powerful advocate for Darwin's theory of evolution, for scientific and technical education in Britain, and for much more. He also married Henrietta Heathorn.

By the time of the steamship, later in the nineteenth century, the trade of empire had been well established along the now-charted inner passage through the reef. From 1890 to the outbreak of World War II, Japanese skippers came to recruit Aboriginal men for labor on the pearl luggers. In March 1899 the entire pearling fleet of the region was wrecked when two cyclones converged on Princess Charlotte Bay where the fleet took shelter. Of the 307 men who died in the disaster, only 12 were "white men."[4] When an old lugger hand from Lockhart River, to the north, commented, "All kind of nation been here before," he was referring to times within living memory of people in the 1970s and 1980s.[5] The federal government of Australia has been the last in a long line to claim ownership and control of the reef.

Archeological evidence indicates that Aboriginal people have occupied Australia for at least 40,000 years. The first people to arrive were ocean-going people, who must have come by canoe or other watercraft that were able to travel some 50 miles across open seas. It is assumed that they came from the islands of Southeast Asia at a time when the sea level was some 200 feet lower than it is today. They landed along the northwest shore of the land that prehistorians call Sahul, the union in one land of Australia, Tasmania, and New Guinea, which came into being when ocean levels fell and joined these land masses to each other. The present-day Gulf of Carpentaria was then landlocked as one of the world's largest lakes.[6] The Great Barrier Reef as we know it did not exist, and a coastal plain up to 125 miles wide extended to a limestone ridge at the shore—the relic of earlier Great Barrier Reefs.[7] The oceans rose and fell again until some 20,000 years ago, a date that marks the beginning of one of the last great rises in sea level to the present day. As the oceans rose, so the corals underneath grew upward to the light. So it was only some 6,000 years ago that the coastline stabilized at roughly its present position. And there is evidence from 5,000 years ago of Aboriginal use of the area of Princess Charlotte Bay.[8]

The human history of the Great Barrier Reef goes back at least as far as 6,000 years ago.[9] But what if people lived here while the ocean rose? Evidence may not be found because coastal dunes and middens may have been washed away.[10] John Campbell says that there could be ancient rock shelters and caves some 65 to 100 feet below present sea level in the Great Barrier Reef, which he views with the eyes of an archeologist. To Campbell the reef is a vast limestone formation honeycombed with cave systems, which, unfortunately for archeologists, happen to be underwater.[11] The idea of the seas rising and inundating inhabited land reminds me of the

French folklore of the submerged cathedral, of bells heard ringing from beneath the sea. Does the Great Barrier Reef have similar drowned places of former human significance? Are there rock shelters similar to those now perilously close to the sea on Stanley Island, for example, that are now underwater? The answer is bound to be *yes*, but the evidence may never be found.

How might these great geological events have been experienced, if that is what happened to the people along these coasts? Perhaps the rising seawaters were experienced as a series of storm surges where the sea washed up over the land, tumbled along with the added force of extreme winds. After the storm was over, after the exceptional surge, perhaps the sea did not retreat but stayed, covering former hunting and reef fishing grounds. The cyclone of 1918, which washed away the Aboriginal mission at Mission Beach, killing five people, is a more recent example of dramatic coastline changes caused by a storm surge of just 10 feet.[12] By 6,000 years ago the coastal fringing lands became islands, with only the tallest hills remaining above water. Now there are 196 of these continental islands. Lizard Island and the Whitsunday Islands were places of retreat as the waters rose.

The sea level stabilized roughly where it is now, and it has been stable ever since—unusually stable, scientists think, because of the greenhouse effect. The good news, says geologist Malcolm McCulloch, is that what he calls the "human greenhouse experiment" of the industrial revolution has taken place at a time when the earth might otherwise have been cooling.[13] In this way of looking at it, events such as greenhouse warming that cause waters to rise may have been countermanded by forces, such as glaciations, that tend to cause waters to fall.

As the seas rose, new coral islands grew from the underlying shelf platform. The Barrier Reef is a mix of two island forms: continental islands and the true coral islands that grow from the reef-building activities of corals. Some islands are remnants; others are new arrivals. Old land is taken; new land rises from beneath. The new coral islands such as Bewick Island, Ingram Island, and Nymph Island were visited by coastal Aboriginal people who left behind, in shell middens, rock tools they brought with them from other places.[14]

The Flinders group of islands are continental islands with reef platforms.

Under the sea there may be relics of life as it was once lived. Ancient middens and sacred sites may lie hidden, undiscoverable, coral encrusted, or disassembled by the action of surf and tide. A reefscape seems pristine, nature in its most "natural," yet the recreational diver may often see relics of shipwreck and anchor damage and can only imagine what more ancient relics may lie concealed by encrustations and sediment. The Whitsunday Islands are one place where definite archeological evidence does exist from Aboriginal occupation dating some 9,000 years ago, and this is from a time before the sea level stabilized.[15]

After the wave of first inhabitants, others came. Many people have passed through the waters of the reef, grateful to leave them alive. Shipwrecks were common, before the advent of modern navigation technologies. Mariners were glad to be rid of the reef, even to face the dangers of Torres Strait to the north. As the early inhabitants of Ship shelter watched for the arrival of strangers, so the strangers were equally apprehensive about encounters with people onshore. Later came the resource raiders, those in transit on the path to wealth through exploitation of the fisheries. Today the reef is well and truly in its tourist period, people transiting for pleasure. This phase will not last forever.

History can be read on the rocks of Stanley Island. The steep red sandstone cliffs bring to mind the red heart of Australia. "The whole aspect," wrote anthropologists Hale and Tindale when they took part in a scientific expedition to the area in 1933, is "one of infertility and aridity."[16] Infertility lies, at this place, in the eye of the early European beholders. When in 1929 the anthropologist Donald F. Thomson traveled east across Cape York from the Gulf of Carpentaria to Princess Charlotte Bay, he recorded the search for food:

> We were almost entirely dependent on what the natives speared, opossum, flying fox, "goanna" or lace lizards and what fish were left in the remaining holes in the river courses. . . . Our course lay roughly east, but actually represented a zigzag from one *nonda* plum tree to another, where the boys picked the yellow fruit until we caught up with them with the horses, when they made off for the next tree in the distance. At first we protested vigorously, but when we had grown accustomed to the astringent flavour of the fruit, we were glad to take our share. Now and then we halted while a "sugar bag" or native bees' nest was cut down.[17]

In addition to food, the three Aboriginal men in Thomson's team found the water. Thomson stayed 12 months with the people in the area of Princess Charlotte Bay. He came as a naturalist and budding anthropologist: "Here, with these happy, genial, carefree fishermen, I served my apprenticeship as an anthropologist."[18] Writing about the experience some 25 years later, he said:

> I remember my first meeting with the Yintjingga as if it were yesterday. It was very early in the morning, and as I approached the sandbar at the mouth of the river, the sea

> was oily calm and intensely blue. The sandbanks against the clear waters of Princes Charlotte Bay and the line of green mangroves behind shone dazzling white. . . . In the shallows was a fisherman armed with multi-pronged fish spear, poised like a statue. . . .[19]

Thomson made detailed photographic and written records of life with the dugong hunters and the rituals associated with hunting and sharing the food. One photograph shows a Yintjingga dugong hunter's grave, which contained parts of the skeletons of some six dugongs. The ribs are heaped high on the outer edge of the grave, where they fan out like the petals of a flower. The skulls are piled on top.

The seas of Princess Charlotte Bay and the Flinders group of islands are rich in shellfish, turtle, fish, and crabs. In front of the entrance to the Endaen rock shelter on Stanley Island, a large midden contains the remains of clams, spider shells, cockles, and faded silvery trochus shells. A rock overhang projects some 25 feet over the area, protecting it from the weather. A curtain of fresh water seeps through from the rock ledge and falls to the rocks below, where it gathers in hollows. The visitor, having climbed the dunes from the beach to Ship shelter, walks on to the Endaen shelter through a freshwater shower into coolness and shade. Here is the Ritz of all shelters. This place had it all.

Researchers Chase and Sutton have studied the Aboriginal languages of Princess Charlotte Bay and the Flinders group of islands. One striking feature they discovered is the way in which a speaker will usually tell what relation in space one person bears to another. Instead of saying "the man became angry," it is "the man (in the south) became angry."[20] Though the linguists are cautious, they believe this reflects the importance of the sea and navigation across it to the people who lived there, who were frequently at sea in dugout canoes. Always knowing where you are in relation to fixed points of reference is crucial in sea crossings. The language of these Saltwater people also evolved distinctive notions of directions, the cardinal directions being given in terms of the prevailing winds. There is a word for each wind from the points of our compass, but, for example, they would translate, variously, as "south wind, cyclone, cold land breeze." The winds relate to compass directions but also to the surrounding land, seascape, and seasons.

The present coastline of Australia has changed in the company of Aboriginal people. Some changes are geological; others lie in the realm of myth. The Flinders group of islands have mythic places. The northernmost island, Clack Island or Ngurromo, is a huge sheer wall of rock jutting out of the sea and surrounded by reefs that make landing difficult even with modern watercraft. It is inhospitable, hard to get to, and has no fresh water, yet it has extensive rock art galleries with many remarkable images.

Clack Island is a place of special significance in myth. Some of the rock art images—the frogs and crabs painted in red ochre and outlined in white—are sorcery figures painted to invoke the powers of the ancestors who first created the islands. According to the creation myth, the island was once the paper-bark hut of two ancestral brothers, Itjipya and Elmbarron, who speared a whale and fled to this place from the whale's relations.[21] The whale became nearby Blackwood Island.[22] Clack Island was a restricted place, open only to adult men of the appropriate kin group. Ancestral brothers guarded the area and were treated with respect. Their powers were invoked through sorcery, song, and painting. Sea creatures feature in the rock art: turtles, starfish, stingrays, jellyfishes, dugongs.

Science tells a story of the formation of continental islands as the waters rose after the last Ice Age. In contrast, Aboriginal people tell a story of origins in which mythic significance is invested in land and reefscape. Science stands detached from daily life. With the right training I might look at the land and read a certain geological and botanical history from it. Myth promises more. If only I knew what to do, what to sing, what to paint, how to invoke the ancestral beings. . . . But they'd have to be *my* ancestral beings; I'd have to be an adult male. I am, however, just a tourist in speculative transit.

Multiple aspects of time and space enter into understanding reefscape. I always get a thrill when I see the pattern of reefs and islands from the air. I take delight in the presumption of the god's eye view. I see the reef as a pattern of islands and rock, and I see a place of history, human habitation, voyaging, and myth. From a plane I see only a small part of the reef, but, with satellite images on the Internet, it is simplicity itself to go higher to more godlike regions still. At the National Geographic website the choice is offered of the view of the reef from 3, 15, to 301 miles above the earth.[23] The huge reef is visible as one system but still as mysterious as ever. How is it possible, how can we ever know, if a crown-of-thorns starfish infestation at the far northern reaches will affect marine life in the far south? See the Pacific Ocean on the satellite image and scroll over to the coast of South America. Imagine the interacting ocean-atmosphere systems of this region and how what happens off the coast of South America has multiple effects in Australia. I see Stanley Island in the 3-mile image and marvel at the powers of the desktop computer.

Stanley Island is, in 1999, part of Lakefield National Park. A walking track leads from a small sandy beach set in the mangroves of Owen Channel up over the hill to the seaward side of the island.

The coastal bush is rich in edible plants, with some 37 species listed in the local region.[24] The beach ridge supports a natural orchard of fruit trees: the reddish-brown wongai plum *Manilkara kauki*, the sweet cherry *Eugenia reinwardtiana*, the small purple fruit of *Terminalia muelleri*, and the shiny purple-black fruit of *Pouteria sericea*. These trees bear fruit at different times of the year and provide an ongoing if small-scale supply. The plants in this fore-dune area have kept their own rhythms in time. Here the mutual influence of people and plants may be read into the landscape. Middens provided organic enrichment of the soil. New plants grew from seeds discarded near shelters. Changes in the distribution of food species have occurred in the company of the people who gathered and used them.[25] Today the visitor walks the interpretative trail through the area, reading the signs to gain insight into a way of life that no longer exists in this particular place.

A large red sandstone cliff stands on the water's edge at the end of the beach, dividing the beach area along which we walked from the mangrove area behind. On one side the sea laps at its base. You can't walk all the way round it. Signs tell us this is a mythological site of the turtle. The walk we tourists take through the national park is a walk through time. The national park signs tell visitors they can land and go as they please, provided they obey the park rules and do no damage. The people who once tended this "park" are gone, lost to time. When Europeans arrived in the 1880s, they brought with them their diseases that decimated the people who had lived here for millennia. Fifty years later the Indigenous population had dropped from 40 to 50 to 15. The remaining people were taken away to camps on the mainland in 1930. Their absence haunts the island. We walk along an elegant boardwalk that takes us above the fragile midden shells around rock art sites with minimal impact. The day is hot and humid, and the protective varnish on the boardwalk sticks to our shoes and squelches underneath.

Histories come together in this place. Tourists are very late arrivals. In the nineteenth century the industries of *bêche de mer* fishing and pearl shelling and the abuses of forced Aboriginal labor compelled the Queensland government to act to control what was going on in the far north. The boundaries of north Queensland were changed to include most of Torres Strait.[26] Missions were established along the coast, and the coastal Aboriginal people were moved to them in part so that Aboriginal labor on the luggers could be supervised at least nominally by mission superintendents. Today pearl shelling has all but vanished as an industry. The reason the reef is not Indonesian or Japanese is largely an accident of history. The Macassan trepang fishermen of some 400 years ago came to raid the resources, not to settle. The influence of the British Empire has waned, and the Japanese did not succeed in invading and conquering Australia in World War II. To the north, Papua New Guinea became self-governing in 1973. To the west, on the other half of its island, Indonesian Irian Jaya has a policy of forced transmigration of people from other parts of Indonesia. The changes to the near north of the reef in the next 50 years will be significant, but for now the reef has escaped the fate of being divided up among neighboring and invading nations. It is not part Japan, part Indonesia, part Britain, part Papua New Guinea, though at times people from each place may well have acted as if they owned it. It is an Australian reef and the largest reef in the world under the management of one nation.

From Stanley Island the task is to make sense of possible futures for the reef. The rock art will fade, as it already has since the early years of the twentieth century. It is too close to the wind and the waves. By 1999 the land and marine rights of Indigenous people achieved heightened significance, with claims for native title over the sea in the area to the north, including islands of the Great Barrier Reef and Torres Strait. Commercial fisheries in the region may have to enter into negotiations with Aboriginal people, rather than automatically assume they have the right to fish. Both Indigenous and non-Indigenous people are taking part in the reinterpretation of "sea space" along the Australian coastline.

For 200 years the conqueror's view has been that Cape York is a land ripe for transformation by capital and empire. Yet still it endures, this seascape north of Cooktown, as land where no one has ever built a city. I doubt whether there is a comparable coastline anywhere else in the world so bereft of urbanization. Rainforest tumbles down to mangroves at the water's edge. Crocodiles lurk in the mangrove mud of the creeks. Aboriginal people have lived here for at least 6,000 years. Once their lack of cities was interpreted as a deficiency; now it is seen as a way of living with the land and sea that was admirable custodianship of the environment. Histories are imperfectly and unequally glimpsed. Alexandria, Rome, Carthage, their rise to power over the Mediterranean and their decline are well charted in the records; long before these cities rose and fell, Aboriginal people traveled the coasts of Cape York. The traces of their history are more scattered than the histories of Europe. They are glimpsed in the archeological record and in common elements of myths of the heroes shared by peoples along the coastline and on the islands to the north.

In Ship shelter on Stanley Island there is an outline of a figure standing upright in a canoe. When I stood in that place and saw this image I remembered the Tagai myth of the Torres Strait Islanders. There the Southern Cross constellation is known as the stars of Tagai. The ancestral hero Tagai stands in a canoe. In his left hand he holds a fishing spear, and in his right hand he holds the fruit *Eugenia*.[27] The Torres Strait Islanders and Cape York Aboriginal people who hold elements of this hero myth in common see in these stars a story of sea heroes and their journeys. Once the heroes sailed the seas. After they became stars, they signal the changing seasons. They govern the large world of the remote heavens and bring order to the everyday lives of the people below. Tagai is a god standing in his canoe, arms outstretched. To me he is like the figure in the Ship shelter, but that is a large leap for a mere visitor to make. Underneath the canoe is an image of a very large crocodile, several times as long again as the canoe. The crocodile is one image of the Cape York culture hero I'wai. I get a sense of the mysteries in this place.

Parts of the Great Barrier Reef have grown in recent geological time, but it is not the only reef there is or has been. In geological history, conditions have favored reef creation at intervals of some 200 million years. There were reefs 400 million years ago and again 200 million years ago, reefs that were created from different balances of different marine life. What makes the Great Barrier Reef special is that it has existed for so long in relation to people. The monumental scale of geological time for once approaches the personally significant, here at Stanley Island. Geological time—400 million years ago, 200 million years ago—events in these times seem too remote to matter, yet they have and will affect people. Reefs in their creation and demise are indicators of climate changes, changes that have made human life possible, good, and may even again one day unbearable. Here the onset of a glacial period destroys corals; there global warming allows them to flourish, to a certain point. Geological time is both deeply alien to the personal subjective feeling of time and also, at this place, deeply relevant to it.

# Stressed-Out Reefscape

Two major threats to the corals of the Great Barrier Reef have emerged over the past few decades: first, during the late 1960s and 1970s, large outbreaks of crown-of-thorns starfishes suddenly began to kill corals; then in 1998 came a more serious threat—coral bleaching. Both events brought into sharp focus the effects that humans are having on the reef. Both events command reflection on the "natural" element in reefscape. I go to the reef, imagining I am having an unparalleled experience of nature. If what I find when I submerge are stark white skeletons of dead corals, I start seeing reefscape anew, this time in relation to modern industrial processes. When the list of possible culprits is examined, the heart sinks. All or some of the following may cause coral destruction: global warming, overfishing, coastal development projects, increased farming along the coast, with associated nutrient runoff and increased sedimentation from soil erosion. The health of coral reefs is affected by just about everything humans are doing.

Why are these disasters occurring now and seemingly compounding in a double dose of destruction? Are the causes natural or is human interference a serious factor? The word "anthropogenic" is used for disturbances caused by human activity. The term "coral stress" is used to indicate the effects of coral diseases or coral predators. The question is one of examining the extent to which these occurrences are "natural" and hence quite likely outside human control or to what extent they are anthropogenic, hence potentially fixable.

In 1962 the crown-of-thorns starfishes were first noticed in large numbers on the reefs of Green Island near Cairns. The crown-sized spiky reddish-green creatures are easy enough to see as they drape themselves over the corals. What first attracted attention was their capacity to sting: their spines are toxic. Coastal Aboriginal people have long known about this danger, from times when the starfishes were rare. At Green Island in the early 1960s when the number of starfishes suddenly increased, reef walkers reported some painful encounters. If a starfish spine pricks the skin, the spine may break off and leave a piece embedded. The wound may not hurt at first, but after a day or two there is painful itching and swelling. If several spines penetrate, vomiting and severe pain follow. At first the starfish was just another reef "stinger" to avoid. The danger it posed to coral emerged later.

By 1962 the numbers of starfishes had increased dramatically. Green Island was particularly hard hit. One good place to observe what was happening was at the underwater observatory created by engineer Vince Vlassoff in 1955. With the aim of sharing his love of underwater life with others, Vlassoff turned an old diving bell left behind after World War II into an observatory. It was sunk to the ocean floor off the end of the Green Island jetty. Here for the first time in Australia, people could climb down and experience reefscape for themselves in observatory conditions. They came in droves.

The observatory was also an excellent place for naturalists to observe the behavior of reef inhabitants.[1] In 1962 a crown-of-thorns starfish took up residence on a patch of coral in plain view from the window. Doctor-cum-naturalist Jack Barnes reported that, far from "resting" on the corals, as was first thought, the starfish was actively destroying them.[2] When the starfish moved, it left behind a patch of dead coral its own size. Barnes discovered that the starfish extruded its stomach over the coral and exuded digestive juices that killed the coral polyps by liquefying them to a green slime. After ingesting the polyps, the starfish moved on to the next feeding place to repeat the process. By the end of 1964, 80 percent of the coral cover at Green Island had been destroyed. It was a patchy piecemeal type of destruction, more than the expanding circle of devastation. Some patches of coral escaped the starfishes and survived to provide a source for later reef regeneration.

The starfishes had arrived. Where they came from was unknown. As information about the starfishes grew, so did the alarm. The adult crown-of-thorns consumes as much as its own body diameter in coral in 24 hours. Each mature female may produce up to 100 million eggs in a spawning season. The implications were horrendous. Starfishes soon started appearing in large numbers farther down the coast from where they were first noticed. Reefs in the Cairns region are discontinuous, with miles between them. Adult starfishes will not move across this gap. Obviously the larvae of the crown-of-thorns were being flushed away and carried to other reefs in the Barrier Reef system. The only good news seemed to be that, even though larvae did get from some reefs to other reefs, they weren't hitting every reef at the same time.

At this early stage in 1962 the larval and juvenile stages of crown-of-thorns had not yet been identified. Research began immediately, well financed by anxious state and federal governments. At James Cook University, John Lucas and Julie Henderson were first to rear the crown-of-thorns larvae to the juvenile stage in an aquarium.[3] This meant that all the intermediate stages of the creature could be identified, so that the early stages of an outbreak could be spotted out on the reef. Under the controlled conditions of their experimental aquaria, Lucas and Henderson varied and monitored the water salinity and temperature. They found that starfishes prefer conditions of reduced salinity and higher-than-average seawater temperatures. Also, by observing the behavior of the starfish larvae, Lucas and Henderson showed that in the first six months of life the starfishes lead a fairly benign life as algae eaters, emerging at night from hiding places in the coral rubble to feed on the microscopic algae on the reef flat. Only in later stages of development do they switch to eating coral in daylight hours, and at this stage they are immediately obvious.

Armed with this knowledge, a program of monitoring for crown-of-thorns outbreaks was established on the reef. An outbreak is defined as occurring when the adult population is large enough that it consumes coral tissue more quickly than the corals can grow. Today the reef is monitored for all stages of the life cycle of crown-of-thorns. There is a pattern and a rhythm to the outbreaks. From Cairns they spread south for the next three to five years to the Innisfail region, then another five to eight years to get to reefs off Townsville, and 10 to 12 years later to the Whitsundays. By this time the Green Island area starts to recover again.[4] There were outbreaks between 1962 and 1976 and between 1979 and 1991. A gap of 30 years between outbreaks may be sustainable, but outbreaks occurring closer together may spell deep trouble for corals already in difficulty from other stresses. That is why researchers are worried about the latest wave of crown-of-thorns, which is following on too swiftly from before. Late in 1999 large numbers of small juvenile starfishes were found in reefs offshore from Cairns to Port Douglas. These may prove the advance guard of another major outbreak one year later.

Outbreaks have happened in the past, before human impact on the reef with intensive agriculture along the coast. In farms far from the coastline, farmers are cutting down trees along riverbanks. Erosion occurs, and soil and fertilizer are washed out to sea in dark plumes of mud visible from the air after rain, increasing the nutrient load in the water. Normally, so many starfish larvae are produced that most will starve. Nutrient-rich waters, from whatever cause, enable larvae to survive in larger numbers and from time to time reach plague proportions. The larvae prefer conditions of high rainfall, when the salinity of the water is reduced. And they like an ocean temperature that is warmer than usual.

If I were writing another book I might follow both stories of ecological catastrophe—the starfish outbreaks and the coral bleaching—from their origins through the fierce scientific debates on the causes of the crises, to consideration of how they should be handled.[5] Instead, I want to consider both forms of stress on corals in the reefscape from the perspective of how people feel about changes to the reef—the ways in which we come to terms with it, the ways we explain it to ourselves, the meaning it has for us. For example, bushfires have a regular devastating effect on the Australian landscape, and this is a way to imagine what happens with massive coral death. "It's like a fire going through," was the way John Lucas described it, when in the 1960s he first saw what was happening on the reefs near Townsville. "It burns and takes a long time for it to get back again. You have to give it time."[6] The stark white skeletons of corals stand after the disaster like the ash-gray and blackened tree trunks in the Australian bush. A new cycle of destruction is charted, this time on underwater landscape. Bushfires, paradoxically, bring renewal to the land and are necessary for the germination of some native species of plants. There may be similar, but different, effects after starfish devastation, where what returns afterward might not be what we consider as beautiful as what has been destroyed. The first to recolonize are the soft corals and seaweeds, and aesthetically we seem to prefer the reef as it was before, with more of the hard corals. But this personal preference may be misguided. According to some views, the feeding starfishes may help to maintain high species diversity on coral reefs, where the death of some individuals, or colonies, allows other species the space they need to grow.

Bleached and dried skeletons of corals have their own beauty and have long been collected and used for decoration, from dining room mantelpieces to the window displays of old-fashioned fish-and-chip shops. Dead coral is viewed somewhat differently today. Coral collecting of live or dead specimens is now forbidden on the Great Barrier Reef. Instead of admiring the intriguing and delicate structure of dead coral branches, perhaps we now think more about what a piece of coral means. Something that was once living has been destroyed for our pleasure. Something now in danger has been carelessly treated as an object for casual consumption. There is a sense that the coral looked better where it used to be, either living or cast up as a skeleton on a coral beach somewhere a long way away.

When I first saw a crown-of-thorns starfish underwater, my first thought was a surprised "Oh, that's what it's like." Then came the frisson of primeval fear. "The enemy is before me." The spiky destroyer lay there, doing what it does superbly well—surviving. Here was the starfish that turns beautiful coral polyps into "green slime"—words that invoke the slime of the alien in horror movies or the primitive wildness of the primordial soup from which life on earth first came. Sight of the starfish inspires the fear that humans may destroy what they love. Here, in this starfish or there in that patch of coral bleaching are the consequences of decisions by governments and transnational corporations to boost industry; here is the result of the personal decision by a farmer to clear trees along a river bank; here is the outcome of the vast natural forces that precipitate climate change with or without human input; here is where tropical storms have rapidly diluted the sea. All these factors, and more, are implicated in crown-of-thorns outbreaks. The starfish glimpsed quietly feeding is plugged into a network of natural and human-induced changes that exceed its comprehension and mine. Nature operates on a scale in which many of the species that have ever lived are now extinct. Which of us will last the longest on earth, as a species? My bet is on starfishes.

For the moment, though, the species to which I belong values the living reef and would like to act to preserve it. Following the horrors of the first crown-of-thorns outbreak, when research money became readily available, it became clear that there was need for some kind of research coordination and monitoring body. The Great Barrier Reef Marine Park Authority, or GBRMPA (commonly pronounced with a growl as "g-broompah"), was proposed in 1975 when legislation created the Great Barrier Reef as a marine park. GBRMPA is a statutory authority that regulates all recreational, building, business, and research activities for the reef. Created in response to the potential disaster of the crown-of-thorns, it is now an influential authority charged with the task of monitoring and managing the Great Barrier Reef as an entire reef system. It maintains a COTSWATCH (crown-of-thorns starfish watch) program, with members of the public invited to write in with reported sightings.

In general, reef managers don't advocate trying to control the starfishes by killing them, as this has proved useless except for very small areas of about 1 hectare. The scale of the problem in an outbreak is too great. In the early 1960s at Green Island divers removed 27,000 adults over a period of 15 months from a 2-hectare patch, but still the island suffered great damage. Now at some popular tourist areas some control is managed by means of an intensive process in which each adult is killed individually, with an injection of sodium bisulfate.

A more workable management strategy will come from studying the larger picture, both the life cycles of the starfish and how humans might be contributing unwittingly to the plague. In time, given no action at all, the starfish numbers will decline, if only as part of the natural cycle of checks and balances in which any organism in plague proportions soon eats all its food and starves. That has happened in the past, with evidence in the fossil record of past outbreaks and recoveries. Past conditions of life were different, and reef managers are keen to act to keep the starfish population within the capacity of the reef to sustain it. Because there was recovery in the past does not mean there will be recovery now, with all the additional stresses humans have placed on the reef.

The 1998 International Year of the Oceans was the hottest year on record for a thousand years; it was also the year of death of corals on a scale never seen before, through coral bleaching. In bleaching, coral loses its color and much more. The brilliant colors of corals come from tiny single-celled algae, the zooxanthellae or symbiotic dinoflagellates, which live in the tissues of the corals in great numbers: between 6 million and 12 million organisms to a square inch of coral tissue. When the zooxanthellae get stressed, they collect together in the hollow column of the coral polyp and leave their host. They bail out into the ocean to take up an independent life. The coral skeleton becomes visible through the now-transparent polyp. Normally, the algae live in symbiosis with the polyps and produce up to 60 percent of their energy: when they bail out, ill health and often death of corals follow.

Coral bleaching is relatively new as a cause of massive death and destruction of corals. It was observed on the Great Barrier Reef before 1998 but on isolated reefs in isolated patches. There have been five similar if less serious outbreaks since 1979, though, curiously, few reports before then. The bleaching effect was linked to a number of causes: both increased sea temperature and decreased water salinity caused by heavy rain can prompt corals to release large numbers of algae. Laboratory studies in the 1980s showed that other stresses such as increased ultraviolet light, sedimentation, and toxic chemicals also may cause bleaching.[7]

The 1997–1998 coral bleaching event was remarkable in its global scope. What happened on the Great Barrier Reef also happened in a linked chain of events across the world. Starting in the Galapagos Islands off the coast of Ecuador, on December 26, 1997, the bleaching passed across the Pacific to arrive consecutively at Fiji, Australia, Southeast Asia, Saudi Arabia, Israel, and lastly, by September 1998, at reefs in the Caribbean. Another remarkable thing about this event was its predictability. Unlike the first crown-of-thorns outbreak, which caught people by surprise, scientists were able to make grim predictions about when coral bleaching would next occur in this worldwide chain of bleaching, and they were proved horribly right.

While the coral reefs were bleaching, the event was monitored by researchers all around the world, who were sitting at home and watching the story unfold on the Internet. Regular real-time reports were posted at a website maintained by the National Oceanic and Atmospheric Administration of the U.S. Department of

Commerce.[8] NOAA uses remote sensing tools in satellites above the tropical Pacific to measure and transmit daily information about sea surface temperature, currents, and wind conditions.

In 1998 this information about sea surface temperature became crucially important for corals. NOAA monitors El Niño events. Periods of unusually warm waters off the equatorial coast of South America happen from time to time as a result of disruptions to the normal interactions between the ocean and the atmosphere. When an El Niño begins, the patterns of winds and ocean upwellings change, and these changes affect weather around the world. As a result, droughts may occur in Australia or floods may occur in California. At the start of 1998, unusually warm waters spread from the coast of South America across the Pacific. On the NOAA website, colorful computer graphics charted the daily progress of the warm water toward Australia and the Great Barrier Reef.

For the corals a small rise in sea surface temperature led to dramatic effects. The wave of warm water passed around the globe, and where it passed, one month later, in many places (but not everywhere) the bleaching began. Month by month the website showed images of the hot spots passing westward across the Pacific Ocean from South America to Australia. Month by month, one month later, many corals bleached and some died. Predictions from the website became grimly accurate. One month the hot spot: the next month bleached corals.

The NOAA website maintained a link to the Global Reef Monitoring Network, where the story was pursued further.[9] Clive Wilkinson is coordinator of the Global Reef Monitoring Network, his job made easier through a NOAA coral e-mail list of observers in all the tropical regions of the world. They e-mailed to the list, and their individual contributions helped build a comprehensive picture of the global extent of bleaching. The Global Coral Reef Network prides itself on being this kind of bottom-up network. One of the reasons for the intense interest in the coral bleaching story, Clive Wilkinson says, is the ready availability of the real-time NOAA reports. The wave of heat passes around the world. People in remote places report the effects. From elsewhere in the world people can log on to see what is happening, as it is happening.

In Australia the Great Barrier Reef Marine Park Authority has been busy evaluating what coral bleaching means for the reef. In 1998 it regularly issued press releases, like this one:

> *Media Release*
> *Subject:*
> *GBRMPA Media Release*
> *Date: Thu, 23 Apr 1998 10:31:37 +1000 (EST)*
> *From: GBRMPA-Media-list@gbrmpa.gov.au*
>
> *Latest on Coral Bleaching—Survey Results*
> *for release 23 April 1998*
>
> *Latest comprehensive aerial surveys, carried out by the Great Barrier Reef Marine Park Authority (GBRMPA) show extensive widespread coral bleaching along the reef coast.*
>
> *The surveys indicate that 88% of inshore reefs are bleached to some extent and about 25% are severely bleached (more than 60% of corals estimated to be affected).*
>
> *Inshore areas in the Mackay Capricorn section are most severely affected, while the reefs in the Far Northern section are the least affected.*

*These estimates from aerial surveys are likely to be conservative since this method tends to underestimate the extent and severity of bleaching.*

*Monitoring programs suggest that this event would appear to be the most severe and extensive ever recorded.*

*There are some reports of very high mortality of bleached corals in the Palm Islands north of Townsville; however further surveys will be needed before the fate of the bleached corals on most reefs can be determined.*

*Bleaching has been recorded on at least five previous occasions on the Great Barrier Reef in the last 20 years; however this current event is believed to be the biggest so far.*

This press release summed up the situation as of April 1998. When the aerial survey data were checked later by researchers swimming over some reefs and checking underwater, it was found that the damage was even worse than indicated from the air.

In October 1998 reports of coral bleaching came in from all around Australia to the annual conference of the Australian Coral Reef Society held at Port Douglas. Luke White, a researcher from the Australian Institute of Marine Science, spoke about the bleaching at Scott Reef, an isolated continental shelf atoll about 200 miles off the coast of northwestern Australia. Early in 1998 his team visited the reef for their usual monitoring purposes. When they left the reef all was well. Two months later in April they returned to find 50 percent of the corals bleached and possibly dead. It was a depressing sight. The bleaching was worst on the surface at the reef crest, where nearly all of the corals were bleached. Even at depths of 100 feet, some 80 percent of corals had been affected.

Ove Hoegh-Guldberg, director of the One Tree Island Research Station on the Great Barrier Reef, worked out how bleaching occurs, and for this he was awarded the prestigious Eureka Prize for science in 1999. He studied both heat stress and the effects of an increase in ultraviolet light to outline the chemical reactions involved. The corals start bleaching from the top where they are exposed to light and only later are the lower shaded surfaces affected. Corals with some pigments of their own did better than those whose colors came only from their departing symbionts. Another factor was water salinity, which is reduced after heavy rain and river runoff from the Queensland coast, and sometimes it could be that bacterial infection is implicated.

Peter Harrison, from Southern Cross University, described the situation at Orpheus Island as "depressing." He saw large areas of dead hard corals and soft corals that had fallen apart or had tumbled off their columns. Dead patches of reef soon became covered in algae. He had one item of good news. His work at Lord Howe Island showed that, although there had been heavy bleaching from March to May 1998, up to 80 percent in some areas, by three months later most of the corals had recovered.

Ove Hoegh-Guldberg commented wryly: "It's one thing to die, but if you don't reproduce it's just as bad." Even if the bleached corals appear to recover, they may suffer reduced capacity to reproduce the next generation through spawning. Researcher Kirsten Michalek-Wagner had some evidence from research on soft corals in an aquarium. These were artificially bleached, and their recovery was charted over a period of years. She found that her heavily bleached sample group suffered a heavy reduction in fertility even after two years. Selina Ward charted similar changes in bleaching at Heron Island in 1999. In many cases she found no eggs in the bleached colonies. Scientists will now be studying coral fertility with renewed interest. Of course, corals also reproduce asexually by budding, a reproductive capacity that gives them a second chance at recovery.

In July 1999 Ove Hoegh-Guldberg produced a startling report for Greenpeace Australia in which he established that the 1997–1998 coral bleaching event was the first clear sign of ocean warming resulting from global warming. This report will be considered in the next chapter. It is sufficient for now to say that his conclusions were grim; he links increased greenhouse emissions to the potential destruction of coral reefs worldwide.[10] If the rate of carbon dioxide in the atmosphere doubles between 1990 and 2070, as reef scientist Terry Done reports could happen, this will also affect the rates at which corals take up calcium from the water for their skeletons.[11] If there is too much carbon dioxide in the air, there will also be more underwater. The effect is of a kind of acid rain on corals, resulting in reduced coral growth and strength, a kind of coral osteoporosis.

No wonder corals get stressed out, with so much happening to them, so fast. Yet, curiously, similar effects are not being found in other marine organisms, and some coral species seem less affected than others. Delicate branching corals like Acropora (the staghorn corals featured heavily in reef photography) are affected much more than robust solid species such as Porites or brain corals. If and when reefs recover, the balance of coral species will inevitably change. The good news is that by August 1999 many of the bleached reefs along the Great Barrier Reef system seemed to be recovering. Even in badly bleached areas some hardy species of hard corals survived. Scientists at the Australian Institute of Marine Sciences in Townsville are taking a cautiously optimistic view that coral reefs have a species diversity that gives a necessary resilience.[12]

Look at it from the coral's point of view. "Stress on corals." Not wussy executive-type stress, easily fixed by on-the-spot massage. This is your life-and-death stress situation. Here are corals that have gone with the flow for 200 million years, and now they're facing the precipitate exodus of their business partners, the zooxanthellae. But this was a deal struck with algae back in the Tethys Ocean. Stick with us, they said; take part in our grand evolutionary strategy, inhabit our extracellular spaces, give us energy from your plant-style photosynthesis; in return, we'll give you—well, what did the corals have to offer? Sanctuary from predation and the safety of a home.

Home is no longer home sweet home. The partnering of coral and zooxanthellae was once a powerful evolutionary achievement: between them they built the reef. And now, between them, they may just let it go, dust to dust, coral reefs to their constituent parts. Carbon dioxide and fine coral sand, that's what will be left. Reefs come and reefs go—that's one lesson we've been learning from earth history. Once the Gogo fields of northwestern Australia were underwater; now they're high and dry and desert. The Gogo reef was a different kind of reef, no symbiosis between polyp and algae, more a community of sponges and sea mosses that formed hard skeletons. Oceans come and oceans go. Symbiotic algae: now you see them, now you don't. Once they cut a deal with the corals, but they didn't mean forever. Deals come apart and circumstances change. The earth gets warmer and corals get a dose of stress.

There is an important lesson to be learned in the long-term view, from the coral fossil record. Corals are good predictors of mass extinction events. When corals get stressed, it has sometimes signaled the coming of a major environmental crisis about a million years or so later (corals work on more expansive timescales than people do). The corals nearly died out with the dinosaurs when a comet slammed into the earth some 65 million years ago. The ancient Greeks believed that comets come when the gods are angry. The medieval view had it that comets were signs of a ruined world that has fallen into sin. I'm more inclined to think of comets as something for which neither the gods nor I am personally responsible, but rather a comet is factor X entering the evolutionary equation. Things trundle along, one thing leads to the next, until one day, suddenly, it doesn't. The sky goes dark and stays that way for years, and nearly all the shallow carbonate-producing animals die with the dinosaurs. *Nearly* all. That qualification is important. A few species survive in a safe nook on some deeper slope, and that is all that is needed to start the corals on the comeback path.

The task is how best to gain knowledge of the limits set by the natural world and work out ways of flowing within those limits. Easier said than done as far as a comet strike is concerned. That's not a *limit*: that's a catastrophe. One large comet will wipe out previous limits and introduce some new ones. The world spins on, with different species coming from underneath, rising from underdog status, to take over those whose survival is no longer so sure. Recall those small ratlike creatures scuttling under the feet of the dinosaurs? They went on to become the ancestors of humans. Questions of intrinsic worth quite simply do not enter. The ammonites were a perfectly decent group of giant shellfish not doing anyone any harm, but something changed dramatically and they became extinct.

That's the story. Going with the flow. A million years before a mass extinction event, the corals vanish, more or less. If this present lot of corals goes, there'll be another mass extinction (us?) and then the corals will come back, give or take another 20 million years. Humans have had it good till now. Now the corals are telling us—beware. Our time may be coming to an end. Watch out for the corals. When they go, we'll follow.

# The Beginning of the End of It All

I once talked with some people who were deeply concerned about an event roughly 4 billion years from now. We were at the 1999 seminar of the Humanity 3000 project, organized by the Foundation for the Future in Seattle, Washington, talking about the year 3000. However, my companions skipped way past the next thousand years to the time when the earth will come to an end, swallowed up by the sun as it explodes in the last stage of its life. This distant event captured their imagination in a way I found strange yet intriguing. Personally, I take the life and death of planetary systems as events way beyond my control. Yet plainly my companions thought differently. For them it came as an absolute human imperative to explore all options for the likely evacuation by humans of the planet: "Otherwise, all this will have been for nothing." By "all this" they meant the vast epic of life on earth, with humans as the latest dominant species. Plainly they assumed that, now that we are here, human life should continue in its dominant role, humans in charge to the end and beyond. Where we have come from determines where we must go: on out into the universe to colonize other solar systems. It was, they resolutely agreed, an absolute imperative. Just as humans have become planet-adapted creatures, so they must move on to become adapted to life in space and on other planets.

As I listened to this conversation with its ultimatum of the "absolute imperative" I found myself in quite some disagreement. It might be their imperative, but it wasn't mine. The order to "do this" or "do that," even when issued in perfectly good faith, tends to provoke a skeptical "says who?" reaction. Perhaps it was because this meeting took place in Seattle, home of space science and computers, that it seemed almost natural to think big in terms of rockets.

This conversation set me thinking about imperatives I might issue about the Great Barrier Reef, assuming I was an absolute dictator. I'd want to be an agreeable benign dictator with the best interests of the reef—and humanity—at heart. Call me Goddess. What would be my ecological imperative? And how would I compel my subjects to listen to me, to do as I said? How would I know I was right? I'd call in reef scientists and listen to them, and by and large their message in 1999 has been one of global glumness. Some kind of imperative might be gleaned from a source such as the Internet forum with reef scientists that followed the Australian Broadcasting Corporation TV program of May 1999 on coral bleaching, *The Silent Sentinels*. The imperative would go something like this: Stop global warming, but it has to stop from *30 years ago*. Because even if all greenhouse emissions stopped now, the effects of past activities will continue into the future. I'd have to be a time-traveling global dictator to get back to the 1960s, throw up wind farms, change the course of world industrialization, and come back to the 1990s to find, perhaps, a new Ice Age instead. Taking to time travel to alter the course of history has its own problems with continuity and being one's own grandma. Everyone's seen *Back to the Future*. I came to the conclusion that even being all-powerful and omnipresent isn't necessarily going to help the cause of preserving the corals, as far as I can tell, or as far as expert scientific opinion, in 1999, can tell. I'd need to be omniscient as well and that is asking way too much, even for a goddess. What this last impossibility illustrates is the difficulty of recognizing both the effects of human actions—for example, global warming, and the limits to human actions.

It took the 1999 Greenpeace report on climate change and the future of the world's coral reefs to bring home the sense of sorrow at the end of planet earth. Written by reef scientist Ove Hoegh-Guldberg, the report examined the extent of the coral bleaching event of 1997–1998 and made forward predictions of widespread coral death and reef destruction over the next 100 years, in the light of computer models of climate change. The conclusions were grim. If greenhouse emissions continue to rise unabated, the end result within 50 years will be catastrophe for tropical marine systems everywhere.[1]

It was in celebration of beauty that I began my year of the reef in 1998. As events unfolded over that momentous year for coral reefs worldwide, I began to feel more like a civil celebrant presiding over a funeral. The invitation is to celebrate the beauty of a place that is doomed. My quest for happiness in reef places becomes instead a eulogy for past delight. Widespread coral bleaching and the subsequent death of corals and their associated reef systems were starting to look like the beginning of the end of it all. If underwater, the diver, the swimmer, and the snorkeller lose themselves in a merging of action and awareness, I wanted to go that one step further, to ask how this way of knowing both oneself and the world anew might survive reentry to land and ordinary life in order to lead to action on behalf of the reef.

One way to engage others in social commitment to preserving the ocean environment is the path of anxiety: to project disaster scenarios for the future of the reef and the planet. Other paths to engagement lie in working through how to put some kind of collective value on the individual experiences of so many reef visitors, to push from the subjective experience of delight to the social dimension of the conservation of natural beauty. Direct experience may lead to concerns of ecological, political, and economic natures, though, equally, it may lead to the desire to own a piece of the real estate. Experience is something, but it is not enough. Science brings knowledge but always with the proviso that what there is to know always exceeds what is known and probably, ultimately, the human capacity to know it. Three things juxtapose here: first, the delight in beauty; second, the imperfection of knowledge; third, the desire to preserve the reef. Living the life that is good for the reef is a problem in knowledge and a problem in social action. Desire encounters conservation policy in the gap between what people say as individuals and what people collectively do.

Port Douglas is a tourist resort in north Queensland, a place where many tourists will come for a tropical holiday with a day trip to the reef as one of the highlights of their stay. In Port Douglas tourism operators talk about "sustainable use" of the reef for the benefit of tourists and tourism—and their jobs. The mayor of Douglas Shire says: "We are a community of marine biologists."[2] Seventy-five percent of employment in the shire is tourism related. The two statements are connected. Young people go to university to study marine biology. They do so in part from a delight in marine life and a desire to know more and in part to have some kind of career in the science. They end up in places they don't expect, as do we all, but in so doing they have the chance to put their knowledge into use in new ways. In Port Douglas marine biology serves tourism, and tourism, reciprocally, is now trying to serve marine biology. Delight morphs into the service of science or the leisure industries and commerce. The task is to place conservation values first in the service of the reef, sustainability over extractive and destructive use. Tourist operators are naturally devastated by the Greenpeace report. If in the worst-case scenario their industry lasts only another 20 years, economic as well as natural disaster is fast looming.

In the early years of the twentieth century, reef workers knocked the reef about in an unthinking, uncaring fashion as if its beauty was an offense. There were few of them then. Now there are a million and more people visiting each year, and soon even remote parts of the reef, two-thirds of all reefs, will be accessible on day trips from the mainland. Something as simple as swimming impacts reef life. Swimmers put on sunscreen lotions that will wash off in the water. There are a million swimmers. Whether sunscreen affects marine life is a necessary subject of study.

A million visitors come to the reef, and marine biology is one of the sciences harnessed both for education and for measuring human impact. Reef scientists turn reef managers. The technical scientific problems morph into "people problems." The new career of "reef manager" was one outcome of the 1975 Commonwealth legislation that established the basis for reef management in the statutory authority of the Great Barrier Reef Marine Park Authority. The GBRMPA took charge partly in response to the crown-of-thorns outbreak but also because of plans by the Bjelke-Petersen Queensland state government in the early 1970s to mine the corals of Ellison Reef for limestone fertilizer and to allow exploration for oil in reef waters. Huge public outrage forced the federal government to intervene, and GBRMPA was created in public recognition of a duty of care toward the reef. It was an admission of the rights of future generations to know and enjoy the reef and to reject plans to sell off its separate parts for fertilizer, food, or fuel. The work of the reef manager embodies the ecological imperative: to keep reef life keeping on. The call is for collective and global action on monitoring and baseline surveys.

Greenpeace, as an environmental activist organization, alerts the public to possibilities given present trends. The Greenpeace report takes the "simulation imperative" of science very seriously, that is, to take present knowledge and push it into the future. Research scientist Ove Hoegh-Guldberg was given this task with coral bleaching, and he gives predictions for the fate of the Great Barrier Reef in 50 years. He creates the scenario on the basis of thorough research but necessarily on the basis of present, not future, knowledge. In the future, scientists will create more proximate simulations, using new research as it becomes available. The "simulative imperative" pushes into an imagined space in the future, in full knowledge that much simulation will prove inaccurate. But it is an "imperative"—do this or else!

In the publicity that followed the release of the Greenpeace report, reef managers called for some caution.[3] The reef is changing, but it is impossible to know how much it is changing unless the status of reef components is monitored over time—for example, the number of species, the abundance of populations, oceanographic and climate information.

The career of reef manager is relatively new. By 1981 the Great Barrier Reef and the islands that lie within it had been inscribed on the UNESCO World Heritage List. World Heritage List status is granted for sites that have exceptional natural and/or cultural values and are important to all nations of the world—the special common heritage of humankind.[4] One of the reef's "outstanding universal values" for the World Heritage List is the fact that it is the single largest system of coral reefs in the world. The

Great Barrier Reef also has the distinction of being the largest area in the World Heritage List.[5] The crucial task is to work out how to manage it for future generations.

There is more to conserving the reef than conserving the corals, difficult as that will be in a world heading toward global warming. The attributes for the World Heritage List include the plants and the birds of cays; the sea grass beds and mangrove communities; the diversity of life forms, from corals to foraminifera to ascidians (sea-squirts) to fleshy algae to dugongs and turtles. It is the reef as a reef system that is valued, indeed, as a beautiful reef system. The duty is to conserve an "area of outstanding natural beauty."

The philosophical traveler will be in accord with the aim of conservation of the places she visits. She may be grateful that there are authorities monitoring her activities as she takes her reef pleasures, though she really wouldn't like to cast herself in the light of being part of the "people problem." Yet to others she is, as she sloshes on the sunscreen, aboard a dive boat that may itself be causing destruction as its anchor chain drags through corals.

Reef life flows on. Reef life ought to flow on. Between one sentence and the next lies the gap between knowing the reef is beautiful and knowing what best to do to keep it that way. Who am I to issue imperatives for what scientists, governments, and reef users and abusers are to do? It is one thing to want reef life to flow on, and I am pleased there are large numbers of people who in a practical everyday sense are taking this duty seriously. It is easy for me to say that there is a debt owed to future generations, so that they too may enjoy this reef, but not so easy for me to do something in practice that will enable this to happen. As an individual I feel powerless in the face of global warming. I live in temperate Melbourne, and when it gets cold I put an extra jumper on to see if I can stave off lighting the gas heater that bit longer, but I know this is not enough to make even a small blip in the trend.

Scuba technology has developed alongside reef sciences, as if describing and explaining reef processes went snorkel in mouth with the desire to keep this patch of nature keeping on. Scientists are attracted to study the reef in part because they find it a beautiful frontier. Scientific knowledge moves from the description of new species to the discoveries of their relationships to each other, to knowledge of reef and oceanic systems with their states both of equilibrium and instability. Along this continuum of knowledge, it is impossible to say where facts leave off and pleasure and evaluation kick in. Reef scientists act as if they want to keep reef life healthy. Like the tourism operators, they do not want to see the source of their livelihood destroyed. Of course, there is more to their love than money.

In the reef experience, words fail to convey the immediacy of pleasure. American philosopher Cheryl Foster has cleverly articulated the "ambient" aspect to the nature experience as a factor that must be considered along with the narrative.[6] By this she means something like the following: a story about the reef, even in the form of a scientific article, will say what words can say, but words can never say it all. The feeling of "ambience" is never conveyed. By this she means "the feeling of being surrounded by or infused with an enveloping, engaging tactility,"[7] in this case, warm water and brilliant sights. Why I like Cheryl Foster's analysis is that the feeling of "losing oneself" isn't talked about in some kind of mystical, ineffable, fuzzy cloud of unknowing sense but encapsulates something that is part of many people's experience of the reef. Why we consider the reef beautiful does not rely solely on the stories we tell about it but springs from the reef experience and draws on the sensuous as well as the knowing self.

Reef systems are dynamic systems. They would be changing even if people weren't there. Keeping reef systems viable means acknowledging change that may not be, at this moment, beautiful but that may be part of the reef cycle of construction and destruction.

As they go about their practical tasks reef managers take onboard the huge responsibility of organizing what is known about the reef and what it means in terms of conservation. In practical everyday terms many people go fishing. Some 30 to 40 percent of Queenslanders say it is what they do for recreation, for pleasure, and for food. The question is how best to link the technical research on reef-fish spawning and reef-fish longevity with regulations to which fishermen will agree. Reef managers created a system of zoning that regulates activities in different places that are sensitive to different pressures. In response to the issue of human responsibility to reef species, some islands are closed to people at times when birds are breeding. Reef managers mediate between the values of developers and the values of conservationists in an ongoing herky-jerky process in which it seems to both sides that they are taking three steps forward and two steps back. The two poles, the pragmatic and the spiritual, range from "It's my property and I can mine it or sell it or develop it" to "All creation is sublime."[8] According to one view, the reef is a pile of limestone. According to the other, the reef is a miracle in which marine life is unfolding its potential in a series of wondrous, fortuitous events. Reef managers have to manage the differences. They have to be able to speak the languages of the scientist and the fishing industry, the tourist operator and the recreational sailor.

Jamie Oliver is a reef manager from GBRMPA who speaks the language of strategic planning, direct management, self-regulation, and active partnerships.[9] Long-term forecasts look good for reef fish, sea grasses, mangroves, and island plants. Uncertainties abound in forecasts for water quality, microalgae, and sea birds. The crown-of-thorns starfish and coral bleaching are seriously bad news. GBRMPA proposes to apply business performance indicators to the reef, in the shape of status-and-response indicators for ecology. What is the status of a particular reef now, and what do we want it to be? It could be classified as being in the state of "pristine," "preindustrial development," "apparently altered," "modified," "deteriorated," or "destroyed." If "destroyed," nothing can be done; if any of the others, planners may plan to restore it or at least help it up a notch on the status table. Planners may plan. It may then happen, given the goodwill of people or regulations and enforcement. The activity called "natural resource management" comes into being as a means to keep nature as "natural" as is best for the good life of the reef. It means managing people. It means taking into account state and federal interests. Managing the reef is a highly political activity.

Once I thought of the Great Barrier Reef as a world-famous wonder of nature, the huge coral ecosystem off the northeast coast of Australia. What can be more natural than sea and sand and fish and birds and reef? Now I've come to understand the reef in more political terms. In taking a reef and reef waters as a common heritage, the move has been from nature to politics and in politics from state to national to international levels. The notion of reef, the biological entity, folds into the notion of reefscape, with the politics of the reef as one of its multiple meanings. The reef may be a collective reef system, but whose reef is it and where do the boundaries lie?

What should be seen as one biological entity is often carved up according to closely defended notions of state and commonwealth legal rights. Some waters are state controlled; others are controlled by the federal government. At present, in 1999, the boundaries of the Great Barrier Reef Marine Park, as declared by the Australian government between 1979 and 1983, start at the tip of Cape York, move out to Bramble Cay and southward from it, excluding 400 reefs in Torres Strait, and extending all the way 1,550 miles north of Fraser Island. It excludes some inshore areas around coastal towns.

The Great Barrier Reef World Heritage Area is different from the area of the Great Barrier Reef Marine Park and also different from a third entity, the Great Barrier Reef Region. (Nothing in this story of boundaries is simple.) The World Heritage Area, in 1999, is defined by specific limits of longitude and latitude, from the tip of Cape York to just north of Fraser Island. It extends from the low-water mark eastward along the Queensland coast, to a point beyond the continental shelf. It includes islands and waters under both state and federal jurisdiction. It is 7 percent larger than the Great Barrier Reef Marine Park and includes, for example, Magnetic Island off Townsville.[10]

The Great Barrier Reef Region is roughly the same, except it has two specific exclusions. First it does not include any Queensland-owned islands or Queensland estuarine waters, including Hinchinbrook Channel. There is a controversial development project going ahead at Oyster Point on Hinchinbrook Channel. This is a Queensland waterway. The other side of Hinchinbrook Island is under federal jurisdiction in the Marine Park, where such development would not be permitted.

If the boundaries of the coral ecosystem of the reef are considered, it seems only natural to throw in the islands of the Torres Strait, coral islands like Waraber and Masig, volcanic islands with fringing reefs like the Murray Islands—Mer, Waier, and Dauer. English fisheries scientist William Saville-Kent, back in 1893, certainly thought the Torres Strait islands a perfectly natural inclusion, when he published *The Great Barrier Reef of Australia*, as indeed back then they were a part of Queensland. The frontier has been, still is, and no doubt will be highly mobile as it is contested by peoples in Queensland and the Torres Strait.

Within the federal jurisdiction there are further subdivisions of the reef into management zones. This part round One Tree Island is a research zone, solely for science. That part, Northwest Reef, is a pink zone, meaning no entry because it is ecologically sensitive. Here is a replenishment area, where fishing is not allowed; there is a place of seasonal closure when the birds are nesting. Recent boundaries enclose sea grass and dugong protection areas.

Turtles may hatch from a nest on land in the state of Queensland and scramble down to the sea across intertidal areas that are state managed and controlled. At the low-tide mark they will cross over into federal territory and then swim into international waters where they live and grow for some years. When they return for the mating season it may be to federal waters just offshore. Later the females haul themselves back to lay their eggs on a beach under Queensland jurisdiction.[11] Biological entities do not fit easily into human constructions of property and who owns what.

The individual human, or reef fish, is part of a community of interrelated parts. Human community may fit within national or state boundaries. Ecological communities rarely do. Coral reefs are found in over a hundred countries. They have the greatest species diversity of vertebrates (animals with backbones) of any community on earth, largely because of hundreds of species of fish.[12] Symbiotic relationships dominate on the reef—that is, relationships of interaction and cooperation among animals and plants—and these cooperative arrangements contribute to the high level of productivity of the reef. Reefs are so endemic and resilient in evolutionary history that it seems they are a semipermanent feature of ocean life. Reefs have a value above their usefulness to people. The World Heritage concept is an attempt to formulate policies that transcend personal, state, and national boundaries for the preservation of the planet. That is the ideal. How policy trickles down from UNESCO headquarters in Europe to a sand cay on the outer edge of the Great Barrier Reef half a world away is something else entirely.

In practice, preserving the natural world is in contradiction to what is understood as progress. Tourists come looking for the experience of a world at its most natural. Many love what they see, but they are seeing nature from a secure base in technology, whether the high-speed catamaran from the Queensland coast to the reef or the island hotel with its airstrip and helicopter launch pad. The experience of the reef is something on sale alongside the luxury accommodations and the swift means of transportation from working life to timeout. Tourists also expect to be well fed. The technologies of food production on the coast affect the reef. Urbanization requires food to be grown elsewhere, and this must necessarily change what nature may once have been 10,000 years ago before any towns existed.

Thinking back 10,000 years ago raises the question of whether we can distinguish changes that have occurred "naturally" in the patterns of the distribution of plants and animals on the reef. A hypothetical question—what might the changes have been, caused by nature alone, if humans had never come to this place? What would happen to the reef if all humans just left?[13] The reef as a physical place would still be there. Yet I for one want to be able to experience this wildness and doing so in comfort adds to the pleasure. The "natural" on the reef today is different from the natural of 10,000 years ago, inevitably and even without the human presence. Disturbance is what happens. Natural systems change naturally.

There is great sadness in reef circles at the extent of coral bleaching and its link with global warming, a process that will continue even if it happened tomorrow that industries ceased their output of greenhouse gases. As Ove Hoegh-Guldberg says, "Book your holiday now." He gives the reef another 20 years.[14] A worst-case scenario is that one of the first casualties of global warming will be the end of the Great Barrier Reef.

In the history of life on earth reefs have come and gone, together with linked groups of reef species. In the Burgess Shale of the Canadian Rockies fossils were by chance preserved, complete, with traces of flesh as well as bone. In these algal reefs of 500 million years ago swam creatures that are ancestors of species known today, together with others so weird they cannot be placed in modern classifications. Chance determined which survived and which did not. Five hundred million years ago there were no humans to mourn the end of the truly weird or the survival of the lucky few. The privilege and the sorrow may be ours, to bear witness to our reef, perhaps soon to mourn its passing, where once the dinosaurs died unlamented. The life lived in awareness of privilege and its attendant responsibility is one of care for and about people and places and things that can be lost, that will inevitably be lost given another 5,000 million years.

Ten thousand years ago there were no cities, no broad-acre agriculture, no burning of fossil fuels, and only a small population of humans. It is hard to imagine what new forms of ongoing life there may be 10,000 years from now in a world that we might perceive, from this vantage point, as somehow going "beyond nature." Imagine human identity in an age of machines that will be smarter than their human creators. Imagine a multiplanetary society, with space exploration swiftly becoming cheaper. Imagine space-adapted humans or a situation in which what moves off-planet may be small space probes rather than humans, probes so smart they will be able to repair themselves in transit. Imagine the ocean with artificial reefs, underwater junk piles of concrete blocks, old ships, scrap metals covered and cemented together by mineral accretion technology—where low-voltage currents are passed through the ocean and precipitate its minerals onto whatever structures happen to be present.[15] Imagine a world that comes into being when nature becomes technologized beyond even these wild dreams.

Imagine a flow of life from the past through the present to the future, life in the process of time walking, or time slithering, or time swimming. From the bacterium to the elephant, each is a unit of something bigger. Individuals are the carriers of genes, where genes relate the individual to the population, first the population of the species whether bacterium or elephant, and secondly the grand genetic process by which all life on earth has come to be and continues on. Though shortly the location of each gene on the human genome will be known, it will be something but not everything. No one individual human bears the entire human genetic burden but rather contributes to a common pool. This can be taken to mean that the self enlarges its biological concern to move out to care more broadly for the whole stream of life on earth and includes the landscapes over which it flows.

I find myself entranced by the fact that parts of the human genome are identical to parts of the genomes of other organisms. We share something like 99 percent of our DNA with the primates, 90 percent in common with the rat, and no doubt a certain amount in common with reef fish, except no one has yet worked it out. In the human genome there is also DNA identical to those of bacteria and viruses. The bacterium, the virus, and the common rat are all in there, somewhere, in my personal genome. As people survive attacks by viruses and bacteria, they not only survive by conquering the infection, "throwing it off," but also by incorporating part of the invading organism within the human genome. One of the fascinating insights from recent work in immunology and genetics is that gene fragments from organisms that caused previous devastating plagues like the Black Death have been found in the human genome. It makes me wonder, sometimes, what am I? I think there's more layabout cat in me than hyperactive rat, somehow—more sedentary sponge than tear-about shrimp.

It is through the genome or sum total of all the genes in an organism that the individual partakes of something above the individual experience. It is through the genome that all living things share in the genetic process of planetary life.

Ocean currents flow from past to future, indifferent to the human presence. The self blends with the not-self. The human blends with the natural in potential ongoing flow to a more technological future. The present, the past, and the future flow together. Life forms have changed climate and the composition of the planet long before today. Bacteria did it first some 3 billion years ago when they began to produce oxygen. Biological changes have led to global changes in the environment, which in turn have led to new opportunities for biological evolution. This long-term process of change has occurred though the history of life on earth, in an unfolding one-way story.

Australian ecologist Terry Done says that, when he began work on coral reefs in the mid-1970s, reefs were assumed to be both fragile and stable, with a kind of "background stability" against which human-induced changes could be measured.[16] Twenty years later the notion of a "balance" to nature is under fire in ecology. The notion of "balance" implies restoration following disturbance; it implies that the history of evolution was worked to achieve a self-regulating interdependence among specialized species. But what if this thing called "nature" is more unstable than previously imagined? What if largely unpredictable environmental changes create disturbances but not restoration to a previous state of equilibrium? In the 1990s ecologists described 17 different types of "coral communities." Just as plant and animal communities in the Australian landscape are different, with here spinifex grassland, there a rainforested mountain slope, equally in reefscape there are differences. Some of the variables that produce variations in coral cover are wave action and water quality; the numbers of fish and other grazers; disturbances caused by cyclones or crown-of-thorns starfish plagues or development; and the ecological history of a community. Terry Done speaks of "least vulnerable" and "highly vulnerable" communities of corals. Work on identifying the factors responsible for variability is one way in which science contributes to management of the Great Barrier Reef.

The trend is to speak of reefs as dynamic systems. Corals grow, adding structure to a reef over millions of years. Then conditions change, and eroding organisms and forces get the upper hand. Proportions of hard-to-soft corals change. The reef recovers, and the reef degrades.[17] If some coral reefs appear to be degrading in a short period of time, that seems terrible, but where this decline might fit into the much longer timescale of biological and geological processes is unclear.

The Greenpeace report on the future of the Great Barrier Reef plugs into some of humanity's deepest fears—loss of control, the extinction of human and other species. The human species may be contributing to changing the planet so we will no longer be predominant. We are in the process of creating our own more capable successors and whether these will be the super-rat, the cyborg, or forms of coral life without zooxanthellae only time will tell.

# Saltwater Spirituality

The two major tourist destinations in Australia are Uluru and the Great Barrier Reef: the world's largest rock and the world's largest reef. Once pilgrims traveled from one holy shrine to the next along paths across Europe to Rome and Jerusalem. From Uluru to the reef in a week is a modern secular pilgrimage, from one global natural wonder and tourist icon to the next.

At the visitor's center at Uluru there is a message from the Anangu, the Aboriginal custodians of the site. It goes like this: "Tourists see the rock and they see it as an exceptionally large rock. But they should look into the rock and if they could do that they would see the spirit that lives there, and they will know about the rock." In this context "knowing about the rock" is more meaningful than the physical act of conquest in climbing it. Climbing the rock is an ephemeral activity, all too soon over and photographed and done with. Traveling to the rock to "look through the rock" is a spiritual pilgrimage in which the search is for relationship with some kind of meaning beyond the immediate material presence. These could be values conceived as residing in nature or in nature as representative of something still beyond nature, the in-some-way supernatural.

The medieval pilgrimage across Europe placed the pilgrim in relationship to Christian values and ideals. At Uluru the traveler is asked to respect the spiritual values of the Anangu in stories that link together the rock, its spirits, the ancient and the everyday, the ancestors and the living. In pilgrimage the site is the vehicle for prompting the mind to reflect on enduring things. The physical place is a means to a reflective, meditative end. This accords with the usual understanding of spirituality as "attachment or regard for things of the spirit as opposed to material or worldly interests." Spirituality can be interpreted in religious terms—to be human is to be a creature of God, possessed of a soul in addition to body and mind. Or it can be interpreted in a secular fashion, as the human tendency to feel a sense of awe and inspiration, to search for fulfillment, to desire to understand, reflect, and appreciate.[1]

The meaning of pilgrimage, say Victor Turner and Edith Turner, who have made a study of religious pilgrimages in Europe, Mexico, and India, lies in the experience of "flow" it engenders. As in sport, play, diving, or playing a musical instrument, there is in a pilgrimage that same merging of action and awareness as crucial components of the experience. There is the same loss of sense of self, a loss that is experienced in a positive way. The result of this kind of loss is not despair but a state of awareness without being aware one is aware. The self is irrelevant. The pilgrim loses himself/herself in flow.[2]

The literal application of "flow" to swimming, snorkeling, and diving is immediately apparent. Visiting the reef, whether taking a dive, going for a snorkel, or just looking, is changing dwelling place from land to sea. With the dive itself, the journey will not be a long one, at least as measured by the clock or by the air left in the air tank. However swiftly time passes, there is still the sense of sojourn, of staying in a place that is not one's own, of visiting another country, in this case underwater. The expectation is for something completely different. For the diver who leaves the book on spirituality on the dive boat, perhaps the hope is also for some kind of personal transformation. The reef is more than itself. Travel to the reef is travel to somewhere more than the reef.

There is a paradox, so often the essence of matters spiritual. In one way it is a fairly trivial paradox, yet in another way it sums up the nature of the human condition. Even as they yearn for something more, divers are hopelessly enmeshed with the material objects necessary for the quest. The technology of liberation, while clever, is clunky. The snorkel, flippers, and goggles are definitely material objects, as are the air tank, the regulator, and the buoyancy control device. Definitely materialistic in and of itself, yet it must be accepted as the necessary means to another kind of end. Air thrusts its material status on the diver. Air, the substance that surrounds and supports us, unnoticed for the most part, suddenly becomes the number one material priority. Nitrogen achieves extra-important status in diving. On land it is the inert element that the body can easily cope with, without the brain having to give it a second thought. Under the pressure of as little as the 60 feet of the amateur recreational diver's limit, nitrogen is absorbed into human tissues, and the diver must not rise too quickly. Sufficient time must be allowed, whether below the water or above it between dives, for nitrogen to diffuse into the bloodstream and pass to the lungs to be exhaled. Forget the spiritual connotations of breath, *spiritus*, soul, or whatever. Air, normally heavy with metaphor, is here a doubly material object.

Sleep, eat, dive. Sleep, eat, dive, read a book on spirituality. Sleep, eat, dive, meditate. Sleep, eat, dive, photograph something underwater. A certain odd pattern to life is set up on a dive boat. Words are not the recreational diver's strength. Ask the diver what it is that links both diving and the book on spirituality and the subject is avoided. It is too personal. Searching for deeper meaning in pleasure combines hedonism and spirituality in a heady mix.

To this is added the sense of transience, that life is fleeting, which is contrasted with something that is assumed to be enduring. Over millions of years Uluru will erode and the reef will change. Yet human life is still fleeting in comparison to rock or reef. Intimations of mortality are all around. We are in the presence of sharks, in chains of predator and prey. Instead of being scary it's actually all rather enticing, energizing, and in Technicolor.

Although I find myself positively vibrating in cosmic sympathy with the ocean, I know in my head that I can't go down most of the paths of faith I meet along the way. I don't yet know how much farther I can push the boundaries of my personal beliefs. I grew up a member of the Anglican Church and loved the immersion in faith, myth, ritual, and music. But I ceased belief at about the age of 13 when the Reverend Cornish preached a sermon denouncing Darwin's pernicious theory of evolution. This must have been in the year 1953 in a church that I later knew had accommodated itself to Darwin's views some 80 years earlier. Today, I'd quite enjoy hearing a rousing mid-nineteenth-century Anglican sermon denouncing Darwinism in terms of the impossibility of man's relationship to ape. At the age of 13 my reaction was that this preacher has got it wrong, and I don't want to listen to him anymore.

What I experienced until age 13 was then called "being religious" in a conventional sense. Now the term "spirituality" is more widely used to encompass both mainstream Western sets of beliefs and a host of others that would have equally alarmed the Reverend Cornish—animism, New Age syntheses, Neo-Paganism, Buddhism, and other ancient religions, and more. Saltwater spirituality might be seen as selfish retreat, except it is an experience many others share and in that sharing comes a feeling of community with nature and others that takes particular strength from its retreat into beauty. Things could be otherwise, we think. The world doesn't have to consist of one war after the other. There are other worlds than the world of conquest and domination, other worlds to experience in far less destructive ways. There are possibilities of rich connection between the self, the ocean, humans and earth, the cosmos. The complexity and intricacy of reef life transforms, if only for the moment, the individual who enters into it and experiences the relation of part to whole. New ways of thinking about self and nature and self and culture emerge with others who experience similar strong feelings of interconnectedness.

A phrase from the Brahms Requiem sings into my head underwater: "For here we have no abiding city." Underwater we have no abiding city; we are not at home. From this world we must depart to seek the world of air, the world of daily cares, great and trivial concerns. A requiem written by Brahms on the death of his mother distils yearning, bereavement, knowledge that this world is transient—yet so, also, will be his grief. Requiem acknowledges mortality and links the transient to the eternal. There is a similar sense of loss in coming back to daily life from underwater experience, even as I know I must. Life in air goes on, and underwater existence is possible only in short bursts.

Spirituality, in its minimal meaning, indicates the sense that there is something here that cannot be explained—mysteries, uncertainties, doubts, and also awe and inspiration. There is something that transcends my small experience and knowledge of the world. The thoroughgoing skeptic might look around and say:

"Well, of course, it's all those fish, and sharks are awesome because they're bigger than you and carnivorous." But it's not just your personal relationship to these sharks; it's the intricate interrelationships of fish to reef, reef to water, water to earth, and all of it to you. Science is one tool to understanding intricacy and complexity, but there comes that urge to ask, "What's all this underwater intricacy for?"

I looked around for ways in which different people were thinking differently about science and spirituality and found calls from many of the formal religious faiths of the world—Judaism, Christianity, Islam, Buddhism—to reexamine their traditions and "green" their faith in the light of recent science.[3] If the wisdom of God is manifest in the works of creation, humans have a duty of care toward the earth. In some ways I wanted more, in other ways less, than the greening of traditional belief. I have some trouble in saying God, or Goddess, or even Gaia—no longer imagined as the personal deity but rather a nongendered force gathering energy from everywhere in the universe. I am more inclined to think about the grand sweep of the evolutionary epic both personally and impersonally because I reckon the impersonal definitely counts in this story. I found myself attracted to the way Czech poet Miroslav Holub (1923–1998) thought the issue through. Holub was an immunologist and a poet. He also had difficulty calling himself a religious person, but he did see himself as a unit of something bigger. He hesitated to call it "spirit." He was much happier with the word "genome." It is through the genome (that is, the sum total of all the genes in an organism) that the individual partakes of something above the individual experience. It is through the genome that all living things share in the genetic process of the planet. Knowledge, and belief in this process, brought Holub to place himself in the position of a religious individual.[4] He went beyond the "here and now" of his individual life to see it as part of an evolutionary whole that stretches from the past history of life on earth into its future. He talked about the relationship of the part to the whole of life, not in openly religious terms but as a kind of instinct or talent for survival.

In the late nineteenth century scientists tried to pin down the difference between living and nonliving things by talking about a "life force." With the genome, "life force" is rendered perhaps less mysteriously as "life entity," the genome or the entity DNA. As genetics becomes more commercialized, geneticists studying computer printouts seem to be removing themselves farther from nature daily. Holub rejected their detachment in favor of the desire for a feeling of connection, to experience the relation of part to whole in his life's work. In a spirit of humility he placed himself before the genetic process of the planet, with the comment "we are not the aim of the process." This "something bigger," the genome, is pictured as an entity with which scientists cooperate in extending human understanding.

Miroslav Holub read the DNA that humans share with other organisms as "the logical record of an integrated organism's evolutionary drama." In the battle between life and death, a compromise has been reached in which life continues by appropriating its former enemies to its own ends. Genetic takeover is part of the process. In interpreting the history of life on earth in terms of the conquistadorial activities of genomes, Holub found acceptance of the fact that the death of one individual is often necessary for the life of others. He sees tragic death as a precondition of a biological optimism. He imagines a biological or genetic supraconsciousness that is an aspect of the life of the planet as a whole and to which the question of the death of the individual is not central. The biologist finds meaning in accepting responsibility for the planet as a whole and as a whole viable system.

I find a certain pleasure in imagining the genetic process of the planet as something that may lead to the genetic process of the solar system, the galaxy even. Either life exists on other worlds or it does not. If it does not, I imagine humans, as they send their emissaries into space in the form of smart, lightweight, self-organizing, self-repairing space probes, may also be sending some of earth's bacteria into space along with them. It may not be humans who take the step out into space but universal life in the form of our humblest, but most successful, inhabitants.

My views, while not exactly secular, are not particularly in synch with spirituality as it is commonly understood. Yet I find a fascination in learning about other ways of seeing the reef in which the element of other-worldliness comes through strongly and differently. I greatly enjoy the reef poems of Australian poet Mark O'Connor, infused as they are with his Christian beliefs. The Biblical story of creation as given in the book of Genesis makes no mention of coral reefs; the marine life of tropical seas did not feature in the oral traditions of desert-dwelling peoples. O'Connor takes up the challenge to include them. In the beginning, O'Connor writes, God surveys his creation in the Garden of Eden and finds it good, by and large, but not yet complex enough. Something more intricate than a Babylonian walled garden, as traditionally conceived, seems called for: the tropical oceans and their myriad interdependent inhabitants. Creation must have a small coral cay at its center. God works until on the seventh day he rests but not in the traditional fashion:

*. . . and on the seventh he donned mask and snorkel*
*and a pair of bright yellow flippers.*
*And later, the hosts all peered wistfully down*
*through the high safety fence around Heaven*
*and saw God with his favourites finning slowly over the coral*
*in the eternal shape of a gray nurse shark,*
*and they saw that it was very good indeed.*[5]

Forget high Romantic angst and Wordsworthian solemnity. The view from the coral cay is that God has a playful if deadly sense of humor.

The sense of coming out of the self to merge with outside things seems to be a core spiritual emotion. The desire to recover a sense of relation to land, sea, and people is a key element in current fascination with the spiritual traditions of Indigenous peoples. In thinking through issues of Indigenous spirituality and the reef, I've not personally gone out into the field as an anthropologist. Nor can I really think myself into a mindset in which the world is experienced as someone brought up in Indigenous ways of belief. What follows is my attempt to set down what I understand about the customary sea beliefs of Aboriginal people living beside the reef. In part the surge of interest in customary marine rights and tenure comes from the recognition that industrial societies and their practices have caused considerable harm and that other ways might prove more enduring.

Coastal Aboriginal groups around the coast of Northern Australia call themselves the Sandbeach or Saltwater people. They believe that the sea as well as the land relates people to their place of origin, their traditional values, resources, stories, and cultural obligations.[6] Reefs and offshore islands contain story places or mythological sites. Anthropologists have mapped the traditional maritime estates of the Sandbeach peoples from Cooktown to Shelbourne Bay, and these maps are available only to some Aboriginal people and the anthropologists involved.[7] Secret knowledge is a core part of Indigenous spirituality.

The Sandbeach people of Cape York say that before there were Pama (where Pama means Aboriginal person or people) the fish, animals, and birds were all like human beings, like Pama.[8] They were spirit people, called Stories, and they established Aboriginal law on the land they themselves created. Stories came to a place and settled down. They changed into rock and other features of the landscape and seascape and are still there as the story places they named for themselves. The Pama today are descended from the Stories who still live on in their own places in the land and sea. The spirits of the ancestors, the Old People, still live in and on the land and sea in their special places. In death, spirit is restored to its place in landscape. The dead live on as spirits in their land and take an interest in what the living are doing.[9] All the creatures, dugong, crayfish, turtle, and snake that are taken for food are not just food but part of a complex web of significance that links people to place to the plants and animals living in that place.[10]

All along the coast there are special places of great cultural significance to Aboriginal people. Dugongs feature in a dugong story place, as does the crayfish, the snake, the diamond stingray, the turtle, each in its own place. Some places are perilous places, dangerous in their links to wind and storm and capable of causing disaster or illness if treated disrespectfully. Humans and animals, land and sea, past and present are woven together, culture intertwined and never separate from nature. It is an animistic spirituality that provides a moral and ethical code by which to live. Core spiritual values live on in secret knowledge, ceremony, ritual, and initiation practices not open to the outsider.

Between precontact Aboriginal customary practices and the growth of the late-twentieth-century spirituality industry comes the Christian mission enterprises on Cape York Peninsula. The Christian missions were set up to counter the worst excesses of the pearl shell and *bêche de mer* industries: forced labor, rape, and murder. In countering one set of evils other evils were introduced: the forced removal of groups from traditional sea country to inland missions (often by kidnapping the children so that the adults were forced to follow); a succession of authoritarian mission superintendents who agreed on only one thing—that their charges would become Christian no matter what. "Spiritually loathsome" is a term used by novelist Thea Astley in her award-winning novel *The Multiple Effects of Rain Shadow* (1996) as she contrasts the physical beauty of an island mission with the horrors that were perpetrated there in the name of doing good. The novel is set on an island she calls Doebin Island, and the events she recounts have parallels with historical events 60 years ago on Palm Island in the Great Barrier Reef, a place to which mainland Aboriginal people were forcibly taken earlier this century.

In recent years Christian beliefs and practices have moved closer to valuing Aboriginal modes of thought much more positively. Elements from both Christian and Indigenous practices and beliefs have come together. The Biblical epic of the Flood of Noah is related to Aboriginal mythic events. The waters of Baptism have relevance to initiation rites. The bora or initiation ceremony is said to be "like church." The crucifix may be worn as an amulet against the actions of sorcerers.[11]

The philosophical traveler to the reef may seek to reimagine the reef by taking onboard some of the ways of knowing the reef, glimpsed, however imperfectly, from Indigenous cultures. The European way with myth is to be concerned with origins—how the world came to be from something quite different. In the Judeo-Christian tradition, God created Heaven and the earth. European myths settle on archetypal characters and events, stories rich in metaphor and allusion that weave deep meaning from past epics into the activities of everyday life. The myths with which I am familiar from European traditions have certain familiar narrative qualities, but these are not universal. When Cape York Aboriginal people tell family stories, they are about relations, not only among people at this one time and place but also relationships of people to place and to the living things in that place—a story that always folds the past into the present, the people into nature. At places where people belong to named moieties or social groupings where exogamy (or marriage outside the group) is practiced, the two moiety divisions may be symbolized by mythic oppositions between different animal species, such as dugong and wallaby stories.[12] Through particular stories of dugong or wallaby, people knew they were related to particular places.

In my family I can provide a family tree of relations, my grandparents, cousins, and so on, back a few generations and forward. I think of what we mean to each other often in terms given by science—there is a genetic tendency toward blue eyes, fair hair, Irish skin. But I do not go beyond thinking in terms of my affection for my relatives or the physical and genetic similarities that link us. I see my Aunt Isolde in my daughter Amy, the same physical features even down to the bun into which Amy ties her hair as a sign of youthful fashion, while for my Aunt Isolde it was the way she always wore her gray hair into old age. I see I have my mother's hands and my father's feet, a quirk I explain by some odd shuffle-around of genetic bits and pieces.

I do not place members of my family in close relation to plants, animals, and country, nor do I conceptualize what we are all up to in terms of Irish myths and legends. If I were to feel around me the presence of the dead, I'd be worried. To see my family relationships in a context of spirituality is alien to me. I certainly see, though, that family relationships can be imagined differently and indeed more richly if different kinds of links are made from the person to the group to the cosmos.

I think this is part of what Aboriginal leader Patrick Dodson means when he says that aspects of Western understanding are "beautifully confounded" by the Aboriginal world view.[13] For non-Aborigines to approach the task of reimagining reefscape, it may be necessary to first acknowledge a state of beautiful confusion and proceed from there.

I come as a tourist to reef places. I do not know half of what I am looking at. I inhabit another place and time. But I can accord these places the same respect my own group expects for cathedrals, shrines, and cemeteries. If I become "beautifully confounded," that is a good place from which to begin.

Along the continuum of knowledge from Indigenous spirituality to modern science lies ethnobiology. Ethnobiology embraces not only knowledge about fish, to take one example, but also the meaning of fishing in a society that is not the researcher's own. The ethnobiologist comes to "another country" hoping to learn what local people believe about the sea, its inhabitants, and their relationship to both. The subject "biology" as it is studied in schools and universities assumes that there is a body of knowledge that stands independently of the cultures of the world's various peoples. The subject "ethnobiology" affirms the value of different ways of knowing about plants and animals. Ethnobotany, with its emphasis on Indigenous knowledge of plants, is coming to the notice of big drug companies, valued more for its money-making potential than for its cultural meanings.

Marine biologist Robert Johannes has written extensively about Indigenous marine practices in the Pacific and Torres Strait. In 1974 Johannes came from the United States to the Micronesian island archipelago of Palau. He wanted to learn from Palauan fishermen something of what they knew from long and close association with their reefs. While Johannes the science student had been deep in textbooks learning how to classify fish, his Palauan counterpart on the island of Tobi was observing fish behavior in relation to the phases of the moon and the rhythm of the tides and currents. The young Palauan made his own hand spears and fish hooks. He learned on the sea, while Johannes wrote up laboratory practicals and took exams.

Johannes was stunned by what he saw and learned. In his 16 months in the Palau region he said he acquired more information that was new to marine science than in his previous 15 years of conventional research.[14] After a few weeks he realized his ecological theories about reef fish and overharvesting failed to include the local political, cultural, and economic aspects of fishing. He had to open his mind to new ways of knowing about fish, and he found his science changed from a narrow marine biology to an enlarged practice of ethnobiology, a progression he narrated in his charming book on people, fish, fishing, and science, *Words of the Lagoon* (1981).

*Words of the Lagoon* is more than a collection of brilliant yarns about fishing with people who still remembered the old ways.

Johannes came away convinced that some, but not all, traditional fishing practices and systems of marine tenure were sounder ecologically than modern resource-raiding practices. The practices of having closed seasons and areas closed to fishing meant that overfishing was avoided. "Owning" the reef, in systems of customary marine tenure, meant there was control over who fished where and when and for what purpose. Open slather in the manner of the Western notion of "freedom of the seas" was not part of the system.

Johannes wrote on fish behavior and fishing and its cultural meaning. He doesn't use the word "spirituality." Few marine biologists do. The word is complex and abstract and often has more overtones of New Age woo-woo than ecologically enlightened practice. Although his research has led Robert Johannes to praise many traditional fishing practices, he is not indiscriminate in his admiration; some practices he sees as destructive. What he would like to see is a coming together of the best in traditional and scientific knowledge.

Inspired by Johannes, Andrew Smith studied the ethnobiology of the Lockhart River and Hopevale communities of Cape York. For thousands of years coastal Aboriginal people have been fishing in what is now the Great Barrier Reef Marine Park area. Like Johannes, Smith wanted to learn what he could from people who were deeply involved with the marine "resource."[15] As he gathered information, he came to see that the word "resource" meant more than fish, or dugongs, or turtles as biological organisms. Smith found a dimension of meaning beyond his Western scientific traditions, something that linked place, species, and cultural traditions in a complex and different perception of coast and sea, its ownership, and its care. In this way of looking at things, ethnobiology is knowledge that Indigenous people have chosen to share, rather than something the visiting biologist has been clever enough to collect. Andrew Smith argued that it helps to throw new light on the meaning of the word "resource." "Resource" is more than the marine harvest. It embraces all shades of meaning of the term "sustenance," including the spiritual dimension. Smith now works in Palau, putting traditional ecological knowledge into practice in conserving local fisheries, both in revitalizing Palau's traditional marine management systems and in rebuffing incursions from foreign business interests in the live-fish trade for restaurants.[16] Insights from Indigenous ways of knowing are helping to put back together the world and its people, which science and its necessary specializations, and business and its ignorance of nature, have artificially segmented.

From "sense of wonder" at the beauty of reef life it is a small step to the notion of the "sense of the sublime," the sense that here is something so wonderful that it transcends the existence of the individual caught up in this particular time and place. The personal and the cosmic are linked together, big world in the small world, macrocosm in the microcosm, echoes of heavenly places in the underwater world, *spiritus with spiritus mundi*. For those who translate this awe into religious terms, the sense of wonder has an ultimate ground in belief in a God. For many people the sheer intricacies of reef relationships and the grandeur of the structure as a whole point inevitably to a divine authorship. If we seek to understand ourselves through landscape, says biographer and critic Rosemary Hill, to know what we are by means of what we are not, much depends on whether we believe in a divine creator.[17] If we are not believers in some supreme being, if faith no longer is central to our understanding of the world, and many people today place themselves in this category, there is still the desire to glimpse some kind of something that is other than ourselves. Swimmers and divers of whatever religious persuasion or lack of it do have something in common as they enter the underwater world, a desire "to know what we are by means of what we are not." There is more to understanding a reef than merely diving it. The beauty of the reef allows the nonbeliever imaginative space to take the notion of "awe" seriously. I think this is where I now must place myself.

For those who find echoes of heavenly places in nature, the "other-worldly" nature of the reef experience is a glimpse of Paradise. Or it can be more down to this earth, a sudden and overwhelmingly immediate perception of just where one stands on the continuum of life, as if a window suddenly opened onto a hitherto closed part of the mind. The privileged position air dwellers imagine they occupy seems to count for nothing in the watery world. Going underwater is a view from the bottom up.

For humanists, atheists, and religious believers, whatever the ultimate cause, the moral and ethical issues remain. How people ought to live with respect to beauty and intricacy is important no matter what the religious persuasion or lack of it.

In any environmental ethics there are some activities on the reef that have to be ruled out of order, such as the use of coral islands for target practice, as with the Australian Defence Forces at Rattlesnake Island off the coast of Townsville in the Great Barrier Reef; the siting of a nuclear power plant on a reef in Taiwan; the use of Johnston Atoll near Hawaii by the U.S. government for the high-temperature incineration of toxic wastes from chemical weapons; bombing with nuclear devices as with the British tests at Montebello Island off the coast of Western Australia and the French at Muroroa Atoll in the South Pacific. Bombing and dynamiting of reefs is out, as is the use of cyanide for killing fish or stunning them for capture for the live-fish export trade. These are destructive of reef habitats and have destructive potential beyond.

Stories of creation deal with such questions as "How does the world come into existence and for what reason?" Science tells a story of the evolution of corals and reef fish 200 million years ago in the Tethys Ocean. Cape York Indigenous traditions tell it differently, in terms of place and relationships of place to place, of place and spirit, and of living things to people. What will emerge from this in the twenty-first century may be some new sense of responsibility to the sea and reef places of the world. A conservation ethic thought through with respect to land may be extended to "sea country," a valuable way of seeing and knowing that Indigenous culture has generously provided. The history of the universe, the history of the planet, the geological history of the reef long before humans, the coming of reef life then human life to this place—all these different experiences of significance, purpose, and beauty from the lifeless to the highly evolved may feed into a new way of knowing the reef, a way that I'd like to think will prove vital and necessary for its continued existence.

# Notes

## Introduction

1. "Groupes of different Corallines growing on Shells, supposed to make this Appearance on the Retreat of the Sea at a very low Ebbtide," engraving after Charles Brooking by A. Walker in John Ellis, *An Essay Towards the Natural History of the Corallines*, 1755, frontispiece, reprinted in Bernard Smith, *European Vision and the South Pacific*, Yale University Press, New Haven, 1985, p. 105.
2. Roland Barthes, cited in Andrew Bennett and Nicholas Royle, *An Introduction to Literature, Criticism and Theory: Key Critical Concepts*, Prentice Hall, London, 1995, p. 194.
3. Peter Brown, *Augustine of Hippo*, University of California Press, Berkeley, 1969, p. 329.
4. Margaret R. Miles, "Happiness in motion: Desire and delight," in Leroy S. Rouner (ed.), *In Pursuit of Happiness*, University of Notre Dame Press, Notre Dame, Indiana, 1995, pp. 38–56, p. 44.
5. Ibid., p. 51.
6. Barry Tobin, "How the Great Barrier Reef was formed," Australian Institute of Marine Science, 1998, at *http://www.aims.gov.au/pages/research/project-net/reefs/apnet-reefs03.html*
7. Carden C. Wallace, "What can a coral collection teach us about the world's reefs?," conference paper, Australian Coral Reef Society Annual Conference, Port Douglas, 16–19 October 1998.

## DIVING FOR OLDIES

1. Jean Devanny, *Point of Departure*, Carole Ferrier (ed.), UQP, 1986, p. xxviii.
2. The history of diving, at *http://uwsports.ycg.com/referencelibrary/noaa/section01/subsection04.html*
3. Jean Devanny, "Down in a diving bell," *The Australian Journal*, 1 Feb 1952, pp. 12–15.
4. Jean Devanny, "Denizens of the Coral Seas," *The Australian Journal*, 1 March 1952, p. 12.
5. Lawrence Martin, "Scuba diving explained: Questions and answers on physiology and medical aspects of scuba diving," at *http://www.mtsinai.org/pulmonary/books/scuba/section9.html*
6. David Sawatsky, "The aging diver," *Diver*, June 1998, pp. 12–14, at *http://divermag.com/archives/june98/divedoctorjune98.html*
7. Mary Brown, "More women please: How to up the girl factor in your club," *Dive Girl*, at *http://www.divegirl.com/women.html*
8. The Women's Scuba Association, at *http://www.womeninscuba.com/wsa/history/index.html*
9. *Dive Girl*, at *http://www.divegirl.com*
10. Bob Friel, "The virtual round-table: Industry leaders share their vision of how, where and why you'll be diving in the next millennium," *Scubadiving*, Nov–Dec 1997, at *http://www.scubadiving.com/feature/specials/roundtable/*
11. Bellaqua Personal Submersibles, at *http://www.flinet.com/~gulfstream/bell.html*

## FISHNESS

1. Jeremy Tager, "Manipulative research—the conservationist viewpoint," Autumn newsletter, Australian Coral Reef Society, no. 25, May 1997, p. 25.
2. *Nature*, vol. 7, 16 January 1873, p. 209.
3. Stephanie Pain, "Swimming for dear life," *New Scientist*, 13 September 1997, pp. 28–32.
4. Thomas Nagel, "What is it like to be a bat?," *Mortal Questions*, Cambridge University Press, Cambridge, 1979, pp. 165–180.
5. Ibid., p. 168.
6. Michael Menduno, "Exploring the ocean planet," *Scientific American Quarterly*, "The oceans," vol. 9, no. 3, Fall 1998, p. 107.

7. "Sustainably managing Queensland's dwarf minke whale tourism industry," CRC Reef Research Center Newsheet, James Cook University, May 1999, at *http://www.gbrmpa.gov.au/~crcreef/4news/News/news202.html*
8. P. K. Anderson, "Dugong behavior: Observations, extrapolations and speculations," in H. Marsh (ed.), *The Dugong: Proceedings of a Seminar/Workshop*, Zoology Department, James Cook University, Townsville, 1979, p. 105.
9. For a discussion on the idea of "richness," see Holmes Rolston, *Conserving Natural Value*, Columbia University Press, New York, 1994, p. 59.

## GOING WITH THE FLOW

1. David Hannan, video, *Coral Sea Dreaming: An Evolving Balance*, 1992, Wild Releasing Pty Ltd.
2. Robert W. Buddemeier and Robert A. Kinzie III, "Reef science: Asking all the wrong questions in all the wrong places?," *Reef Encounter*, vol. 23, 1998, at *http://www.uncwil.edu/isrs/reef-encounter/re23/reef_encounter23.htm*
3. Stephanie Pain, "Swimming for dear life," *New Scientist*, 13 September 1997, pp. 28–32.
4. Terry Done, "Science for management of the Great Barrier Reef," Australian Institute of Marine Science, 1998, at *http://www.aims.gov.au/pages/research/smgbr/smgbr05.html*
5. Paul L. Jokiel, "Long distance dispersal by rafting: Re-emergence of an old hypothesis," *Endeavour*, New Series, vol. 14, no. 2, 1990, pp. 66–73.
6. Mihaly Csikszentmihalyi, *Beyond Boredom and Anxiety: The Experience of Play in Work and Games*, Jossey-Bass, San Francisco, 1975, p. 36.
7. Charles Sprawson, *Haunts of the Black Masseur: The Swimmer as Hero*, Vintage, London, 1993, p. 13.
8. "Tao te ching," in Frank N. McGill (ed.), *Masterpieces of World Philosophy*, HarperCollins, New York, 1990, p. 113.
9. Lao-tzu, *Tao Te Ching: A Book About the Way and the Power of the Way*, a new English version by Ursula K. Le Guin in collaboration with J. P. Seaton, Shambhala Publications, Boston and London, 1997, p. 125.
10. Sarah Allan, *The Way of Water and Sprouts of Virtue*, State University of New York Press, Albany, 1997, p. 16.
11. Lao-tzu, interpreted by Ursula K. Le Guin, p. 58.
12. Ibid., p. 98.
13. Ibid., p. 110.

14. Ibid., p. 77.
15. John Mulvaney and Johan Kamminga, *Prehistory of Australia*, Allen & Unwin, Sydney, 1999, pp. 116, 173.
16. Peter F. Sale (ed.), *The Ecology of Fishes on Coral Reefs*, Academic Press, San Diego, 1991, p. 11.
17. J. D. Collins and C. C. Wallace, "Coral reefs," in L. S. Hammond and R. N. Synot (eds.), *Marine Biology*, Longman Cheshire, Melbourne, 1994, p. 218.
18. Charles Birkeland (ed.), *Life and Death of Coral Reefs*, Chapman and Hall, New York, 1997, p. 10.
19. Daniel B. Botkin, *Discordant Harmonies: A New Ecology for the Twenty-first Century*, Oxford University Press, New York, 1990, p. 149.
20. Sohail Inayatullah, "Causal layered analysis. Poststructuralism as method," *Futures*, vol. 30, 1998, pp. 815–829.
21. John A. Long, *The Rise of Fishes: 500 Million Years of Evolution*, University of New South Wales Press, Sydney, 1995, p. 100.
22. Ibid., p. 173.
23. J. E. N. Veron, *Corals in Space and Time: The Biogeography and Evolution of the Scleractinia*, University of New South Wales Press, Sydney, 1995, p. 118.
24. J. H. Choat and D. R. Bellwood, "Reef fishes: Their history and evolution," in Peter. F. Sale (ed.), *The Ecology of Fishes on Coral Reefs*, Academic Press, San Diego, 1991, p. 59.
25. Colin W. Stearn and Robert L. Carroll, *Paleontology: The Record of Life*, Wiley, New York, 1989, pp. 288, 309.
26. Connie Barlow, *Green Space, Green Time: The Way of Science*, Copernicus, New York, 1997, p. 229.

## WHEN THE REEF WAS OURS

1. Athol Chase, "All kind of nation," *Aboriginal History*, vol. 5(1), 1981, pp. 7–19 at p. 11.
2. Regina Ganter, *The Pearl-Shellers of Torres Strait: Resource Use, Development and Decline, 1860s–1960s*, Melbourne University Press, 1994, p. 4.
3. Thomas Lowah, *Eded Mer: My Life*, Rams Skull Press, Kuranda, 1988, p. 45.
4. Tape ID408, North Queensland Oral History Project, History Department, James Cook University, Townsville.
5. Tape ID410, North Queensland Oral History Project, History Department, James Cook University, Townsville.

6. Norman Bartlett, *The Pearl Seekers*, Andrew Melrose, London, 1954, p. 226.
7. Arthur C. Clarke, *The Coast of Coral*, Shakespeare Head Press, London, 1956, p. 16.
8. Owen Mass, *Dangerous Waters*, Rigby, Adelaide, 1975, p. 80.
9. Chase, "All kind of nation," p. 14.
10. Regina Ganter, "The Japanese experience of north Queensland's mother of pearl industry," Report submitted to the Great Barrier Reef Marine Park Authority, Townsville, 1998.
11. Lowah, *Eded Mer: My Life*, p. 51.
12. Tape ID182, North Queensland Oral History Project, History Department, James Cook University, Townsville.
13. Ron Edwards, *Traditional Torres Strait Island Cooking*, Rams Skull Press, Kuranda, 1988, p. 55.
14. Lowah, *Eded Mer: My Life*, p. 81.
15. Tape ID183, North Queensland Oral History Project, History Department, James Cook University, Townsville.
16. Interview with Lionel Wickham, oral history tape, Great Barrier Reef Marine Park Authority, Townsville.
17. Barbara Walton, "A colorful reef history," *Sunday Mail*, Queensland, 5 July 1998, p. 52.
18. Tape ID181, North Queensland Oral History Project, History Department, James Cook University, Townsville.
19. Nonie Sharp, "Reimagining sea space: From Grotius to Mabo," in Nicholas Peterson and Bruce Rigsby (eds.), *Customary Marine Tenure in Australia*, Oceania Monograph 48, University of Sydney, Sydney, 1998, p. 49.
20. Dermot Smyth, *Understanding Country: The Importance of Land and Sea in Aboriginal and Torres Strait Islander Societies*, Australian Government Publishing Service, Canberra, 1994, p. 2.
21. *Voices from the Cape. Aboriginal People Sharing Their Views on the Sea Country of the Eastern Cape York Peninsula*, videotape, GBRMPA, 1995.
22. Nonie Sharp, "Reimagining sea space," p. 58.
23. Nonie Sharp, "Croker Island: Marks in the sea," *Arena Magazine*, 36, August–September 1988, p. 8. Nonie Sharp, "Terrestrial and marine space in imagination and social life," *Arena Journal*, 10, 1998, p. 51.
24. A. Chase and P. Sutton, "Hunter-gatherers in a rich environment: Aboriginal coastal exploitation in Cape York Peninsula," in Allen Keast (ed.), *Ecological Biogeography of Australia*, vol. 3, Junk, The Hague, 1981, pp. 1819–1852, p. 1822.

25. John Bradley, "We always look north: Yanyuwa identity and the maritime environment," in Nicholas Peterson and Bruce Rigsby (eds.), *Customary Marine Tenure in Australia*, p. 132.
26. Chase and Sutton, "Hunter-gatherers," p. 1825.
27. Ibid., p. 1840.
28. Michael McKenna, "Prized ocean catch," *Sunday Mail*, Queensland, 12 July 1998, p. 14.
29. Kevin Meade, "Fishing in dire straits," *The Weekend Australian*, 28–29 August 1999, p. 6.
30. "Anglers fear tourist-led takeover of the Great Barrier Reef," *Courier Mail*, 13 August 1998.
31. Chuck Birkeland, "Disposable income in Asia: A new and powerful external pressure against sustainability of coral reef resources on Pacific islands," *Reef Encounter*, vol. 22, at *http://www.uncwil.edu/isrs/reef-encounter/re22/reef_encounter22.htm*
32. "Anglers fear tourist-led takeover," *Courier Mail*, 13 August 1998.

## REEFSCAPE WITH SEA SERPENTS

1. William Saville-Kent, *The Great Barrier Reef of Australia: Its Products and Potentialities*, 1893, facsimile reprint, John Curry O'Neil, Melbourne, 1972, p. 331.
2. Ove Hoegh-Guldberg, *Climate Change, Coral Bleaching and the Future of the World's Coral Reefs*, Greenpeace, 1999, at *http://www.greenpeace.org.au/info/archives/climate/index.html*
3. A. J. Harrison, *Savant of the Australian Seas: William Saville-Kent (1845–1908) and Australian Fisheries*, Tasmanian Historical Research Association, Hobart, 1997, p. 17.
4. Ibid., p. 25.
5. Ibid., frontispiece.
6. Regina Ganter, *The Pearl-Shellers of Torres Strait: Resource Use, Development and Decline, 1860s–1960s*, Melbourne University Press, 1994, p. 170.
7. Saville-Kent, *The Great Barrier Reef of Australia*, p. 214.
8. Ibid., p. 215.
9. Maurice Yonge, *A Year on the Great Barrier Reef*, Putnam, London, 1930, p. 24.
10. Saville-Kent, *The Great Barrier Reef of Australia*, p. 41.
11. Ibid., p. 27.
12. Ibid., p. 24.
13. Ibid., p. 26.

14. Henry Reynolds, *With the White People: The Crucial Role of Aborigines in the Exploration and Development of Australia*, Penguin, Ringwood, 1990, p. 223.
15. A. J. Harrison, p. 7.
16. Robert Johannes, *Words of the Lagoon: Fishing and Marine Lore in the Palau District of Micronesia*, University of California Press, Berkeley, 1981, p. 131.
17. *Land and Water*, 3 January 1891, p. 6.
18. Bernard Heuvelmans, *In the Wake of the Sea-Serpents*, Rupert Hart-Davis, London, 1968, p. 268.
19. *Land and Water*, 25 April 1891, p. 535.
20. Heuvelmans, *In the Wake of the Sea-Serpents*, p. 297.
21. Ron Westrum, "Knowledge about sea-serpents," in Roy Wallis (ed.), *On the Margins of Science: The Social Construction of Rejected Knowledge*, Sociological Review Monographs no. 27, University of Keele, Keele, 1979, pp. 293–314.
22. Heuvelmans, *In the Wake of the Sea-Serpents*, p. 268.
23. Rupert T. Gould, *The Case for the Sea-Serpent*, Singing Tree Press, Detroit, 1969, p. 183.
24. Saville-Kent, *The Great Barrier Reef of Australia*, p. 323.
25. "The watch on Tasmania's sea monster," *The Australian Women's Weekly*, 28 March 1962, pp. 18–19; William Joy, "Sea riddles as old as time itself," *Sunday Telegraph*, 11 March 1962, p. 39.
26. Loren Coleman, "The meaning of cryptozoology," at *http://www.ncf.carleton.ca/~bz050/HomePage.czmean.html*

## THE TEARS OF THE TURTLE

1. C. Limpus, "Marine turtles of the Great Barrier Reef World Heritage Area," *State of the Great Barrier Reef World Heritage Area Workshop*, Great Barrier Reef Marine Park Authority, Townsville, 1997, p. 256.
2. H. Marsh et al., "Conserving marine mammals and reptiles in Australia and Oceania," in C. Moritz and J. Kikkawa (eds.), *Conservation Biology in Australia and Oceania*, Surrey Beatty and Sons, Chipping Norton, 1993, p. 232.
3. Kitty Monkman, "Over and under the Great Barrier Reef," *The Cairns Post*, Cairns, 1975, p. 32.
4. Noel Monkman, *Quest of the Curly Tailed Horses: An Autobiography*, Angus & Robertson, Sydney, 1962, p. 147.
5. William Saville-Kent, *The Naturalist in Australia*, Chapman & Hall, London, 1897, p. 237.

6. Brian King, "Raine Island: Green turtle rookery of world significance," *Australian Natural History*, Winter 1984, p. 183.
7. Ibid., p. 185.
8. Ron and Valerie Taylor, video, *Blue Wilderness*, 1992, Episode 2, Orana Films, National Geographic, and the Australian Broadcasting Company.

## SEA GRASS HARVEST

1. Mike Cappo, Daniel M. Alongi, David McB. Williams, and Norman Duke, *A Review and Synthesis of Australian Fisheries Habitat Research*, Australian Institute of Marine Science, February 1998, at *http://www.environment.gov.au/marine/manuals_reports/afhr/vol2is1.html*
2. Michelle Richards, "Options for introducing BRD's (bycatch reduction devices) onto prawn trawlers in the Great Barrier Reef Marine Park," at *http://cathar.tesag.jcu.edu.au/~tgpjc/TG3201/ Background/BRDs/brd.htm* and *A Guide to Bycatch Reduction in Australian Prawn Trawl Fisheries*, Australian Maritime College, 1997, at *http://www.amc.edu.au/pub/bycatch/index.html*
3. Terry Done, *Science for Management of the Great Barrier Reef*, Australian Institute of Marine Science, 1999, at *http://www.aims.gov.au/pages/research/smgbr/smgbr06.html*
4. GBRMPA media release, 15 January 1999.
5. "Sunfish NQ response to Queensland east coast trawl fishery: Draft management plan, 4 July 1999," at *http://www.sunfish.org.au/news/Trawl_response.htm*
6. CRC Reef Research Center, "Seagrass watchers get scientific about coastal monitoring," *Exploring Reef Science*, August 1998.
7. *The Harper Dictionary of Modern Thought*, Harper & Row, New York, 1988, p. 77.
8. Daniel Pauly, *On the Sex of Fish and the Gender of Scientists*, Chapman & Hall, London, 1994, p. 121.
9. Barbara Walton, "A colorful reef history," *Sunday Mail*, Queensland, 5 July 1998, p. 52.
10. Tape ID184/4, North Queensland Oral History Project, History Department, James Cook University, Townsville.
11. John Campbell, "The role of fishing in Aboriginal society before European arrival in Australia," in A. Smith (ed.), *Workshop on Traditional Knowledge of the Marine Environment in Australia*, Great Barrier Reef Marine Park Authority, Townsville, 1990, p. 3.

12. F. S. Colliver, "Historical aspects of Hinchinbrook Island," *Queensland Naturalist*, vol. 22, 1978, pp. 17–24.
13. "Dugongs in the Great Barrier Reef," CRC Research Center, James Cook University, Townsville, 1998.
14. H. Marsh et al., "Endangered and charismatic megafauna," *The Great Barrier Reef. Science, Use and Management: A National Conference. Proceedings*, Great Barrier Reef Marine Park Authority, Townsville, vol. 1, 1997, pp. 124–138.
15. Joan Crawford, "Dollar value and trends of major direct uses of the Great Barrier Reef Marine Park," *Reef Research*, vol. 8, no. 2, June 1998, p. 1; "Trawling nets $150M," *CRC Reef Research News*, December 1996.
16. H. Marsh and G. B. Rathbun, "Development and application of conventional and satellite radio tracking techniques for studying dugong movements and habitat use," *Australian Wildlife Research*, vol. 17, 1990, pp. 83–100.
17. Tape ID184, 4/5, North Queensland Oral History Project, History Department, James Cook University, Townsville.
18. Alpha Helix Expedition, *Billabong*, 1996, documentary (moving image, mute), National Film and Sound Archive, Canberra.
19. *CRC Reef Research News*, "Fruits of the sea: Seeds of coastal sea grasses," July 1997.
20. *CRC Reef Research News*, "Gardening sea grasses," November 1995.
21. "Farming the sea," interview with Peter Rothlisberg, CSIRO Aquaculture and Biotechnology Program, by Sally Dakis, ABC Ocean Week Feature, at *http://www.abc.net.au/rural/oceanweek/feature2.htm*
22. "Marine bioprospecting for the National Cancer Institute," *Reef Research*, issue 7, 1998, at *http://www.reefnet.org/issue7/research7.html*
23. Amanda Vincent, "Marine species in traditional medicine," *Reef Encounter*, vol. 23, 1998, at *http://www.uncwil.edu/isrs/reef-encounter/re23/reef_encounter23.htm*
24. John Bradley, "We always look north: Yanyuwa identity and the maritime environment," in Nicholas Peterson and Bruce Rigsby (eds.), *Customary Marine Tenure in Australia*, Oceania Monograph 48, University of Sydney, Sydney, 1998, p. 134.
25. Donald F. Thomson, "The hero cult, initiation and totemism on Cape York," *Journal of the Royal Anthropological Institute*, vol. lxiii, July–December 1933, p. 460.

26. Dermot Smyth, "Caring for sea country: Accommodating Indigenous people's interests in maritime protected areas," in Susan Gubbay (ed.), *Marine Protected Areas: Principles and Techniques for Management*, Chapman & Hall, London, 1995, p. 154.
27. Athol Chase, as cited in Andrew Smith, *Usage of Marine Resources by Aboriginal Communities on the East Coast of Cape York Peninsula*, Great Barrier Reef Marine Park Authority, Townsville, 1989, p. 87.
28. Pauly, *On the Sex of Fish and the Gender of Scientists*, p. 121.
29. Ibid., p. 122.

## CHICK CITY

1. Myriam Preker, "Raising boobies," *Australia Nature*, Winter 1999, vol. 26, no. 5, pp. 28–37.
2. Emma Gyuris, "Impact of visitors on the breeding success of the bridled tern, *Sterna anaethetus*, in the Great Barrier Reef," Australian Coral Reef Society, Scientific Conference, Port Douglas, 16–19 October 1998.
3. K. Hulsman, P. O'Neill, T. Stokes, and M. Warnett, "Threats, status, trends and management of seabirds on the Great Barrier Reef," *The Great Barrier Reef: Science, Use and Management*, 1996, at *http://www.reef.crc.org.au/3reports/conference1/volume1/s3_4.html*
4. H. Heatwole, P. O'Neill, M. Jones, and M. Preker, "Long-term population trends of seabirds on the Swain Reefs, Great Barrier Reef," Technical Report No. 12, CRC Reef Research Center, James Cook University, Townsville, 1996.
5. D. R. Stoddart, P. E. Gibbs, and D. Hopley, "Natural history of Raine Island, Great Barrier Reef," *Atoll Research Bulletin*, no. 254, The Smithsonian Institution, 1981, p. 33.
6. Tom Griffiths, *Hunters and Collectors: The Antiquarian Imagination in Australia*, Cambridge University Press, Melbourne, 1996, p. 260.

## AN ISLAND IN TIME

1. Josephine Flood, *The Riches of Ancient Australia: A Journey into Pre-history*, University of Queensland Press, St. Lucia, 1990, p. 127.
2. J. M. Beaton, "Evidence for a coastal occupation time-lag at Princess Charlotte Bay (North Queensland) and implications for coastal colonization and population growth theories for Aboriginal Australia," *Archaeology in Oceania*, vol. 20, 1985, p. 4.

3. J. B. Jukes, *Narrative of the Surveying Voyage of H.M.S. Fly 1842–1846*, T. & W. Boone, London, 1847, p. 106.
4. Norman Bartlett, *The Pearl Seekers*, Andrew Melrose, London, 1954, p. 251.
5. Athol Chase, "All kind of nation," *Aboriginal History*, vol. 5(1), 1981, pp. 7–19.
6. John Mulvaney and Johan Kamminga, *Prehistory of Australia*, Allen & Unwin, Sydney, 1999, p. 113.
7. Ibid., p. 115.
8. John Campbell, "Role of fishing in Aboriginal society before European arrival in Australia," in A. J. Smith (ed.), *Workshop on Traditional Knowledge of the Marine Environment in Northern Australia*, Great Barrier Reef Marine Park Authority, Townsville, 1989, p. 3.
9. J. M. Beaton, "Evidence for a coastal occupation time-lag at Princess Charlotte Bay," p. 1.
10. Roger Cribb, "Sites, people and archaeological information traps; a further transgressive episode from Cape York," *Archaeology in Oceania*, vol. 21, 1986, p. 171.
11. John Campbell, "Role of fishing in Aboriginal society before European arrival in Australia," p. 2.
12. M. J. Rowland, "Population increase, intensification, or a result of preservation? Explaining site distribution patterns on the coast of Queensland," *Australian Aboriginal Studies*, 1989, no. 2, p. 37.
13. Malcolm McCulloch, "Climate records from corals," conference paper, Australian Coral Reef Society, Port Douglas, 17 October 1998.
14. J. B. Beaton, "Archaeology and the Great Barrier Reef," *Philosophical Transactions of the Royal Society of London, Series B*, vol. 284, pp. 141–147.
15. Bryce Barker, "Use and continuity in the customary marine tenure of the Whitsunday Islands," in Nicolas Peterson and Bruce Rigsby (eds.), *Customary Marine Tenure in Australia*, Oceania Monograph 48, University of Sydney, Sydney, 1998, p. 91.
16. H. B. Hale and N. B. Tindale, "Aborigines of Princess Charlotte Bay, North Queensland," *Records of the South Australian Museum*, vol. 5(1), August 1933, p. 64.
17. Donald F. Thomson, "The hero cult, initiation and totemism on Cape York," *Journal of the Royal Anthropological Institute*, vol. lxiii, July–December 1933, p. 456.
18. Donald F. Thomson, "The fishermen and dugong hunters of Princess Charlotte Bay," *Walkabout*, vol. 22, 1956, p. 34.
19. Ibid., p. 35.

20. A. Chase and P. Sutton, "Hunter-gatherers in a rich environment: Aboriginal coastal exploitation in Cape York Peninsula," in Allen Keast (ed.), *Ecological Biogeography of Australia*, vol. 3, 1981, Junk, The Hague, p. 1840.
21. Grahame Walsh, *Australia's Greatest Rock Art*, E. J. Brill, Bathurst, 1988, p. 144.
22. Hale and Tindale, "Aborigines of Princess Charlotte Bay," p. 91.
23. *http://www.nationalgeographic.com/resources/ngo/education/xpeditions/resources.html*
24. R. A. Hynes and A. K. Chase, "Plants, sites and domiculture: Aboriginal influence upon plant communities in Cape York Peninsula," *Archaeology in Oceania*, vol. 17, 1982, p. 41.
25. Ibid., p. 49.
26. Regina Ganter, *The Pearl-Shellers of Torres Strait: Resource Use, Development and Decline 1860s–1960s*, Melbourne University Press, Melbourne, 1994, p. 195.
27. Nonie Sharp, *Stars of Tagai: The Torres Strait Islanders*, Aboriginal Studies Press, Canberra, 1993, p. 4.

## STRESSED-OUT REEFSCAPE

1. Tape 1D184/3, North Queensland Oral History Project, History Department, James Cook University, Townsville.
2. John H. Barnes, "The crown of thorns starfish as a destroyer of coral," *Australian Natural History*, vol. 15, 1966, pp. 257–261.
3. John Lucas, James Cook University Archives, Oral History Collection, 5.5.1995.
4. Great Barrier Reef Marine Park Authority, "Crown-of-thorns starfish on the Great Barrier Reef: The facts." Pamphlet, update March 1997.
5. Jan Sapp, *What Is Natural? Coral Reef Crisis*, Oxford University Press, New York and Oxford, 1999.
6. John Lucas, James Cook University Archives, Oral History Collection, 5.5.1995.
7. Ibid., p. 191.
8. The NOAA website is at *http://www.pmel.noaa.gov/toga-tao/el-nino/nino-home.html*
9. For an overview of the Global Coral Reef Monitoring Network, see *http://coral.aoml.noaa.gov/gcrmn/gcrmn.html* and Clive Wilkinson, "The 1997–1998 mass bleaching event around the world," at *http://coral.aoml.noaa.gov/gcrmn/mass-bleach/html*

10. Ove Hoegh-Guldberg, *Climate Change, Coral Bleaching and the Future of the World's Coral Reefs*, Greenpeace Australia, 1999, at *http://www.greenpeace.org.au/info/archives/climate/index.html*
11. Murray Hogarth, "Reef faces 'acid rain' damage from sea," *Sydney Morning Herald*, 2 May 1998.
12. *CRC Reef Research News*, vol. 6, August 1999, p. 7.

## THE BEGINNING OF THE END OF IT ALL

1. Ove Hoegh-Guldberg, *Climate change, coral bleaching and the future of the world's coral reefs*, Greenpeace Australia, 1999, at *http://www.greenpeace.org.au/info/archives/climate/index.html*
2. Address to the Australian Coral Reef Society, Port Douglas, 17 October 1998.
3. Australian Broadcasting Company TV, *7.30 Report*, 7 July 1999.
4. P. H. C. Lucas, T. Webb, P. S. Valentine, and H. Marsh, *The Outstanding Universal Value of the Great Barrier Reef World Heritage Area*, Great Barrier Reef Marine Park Authority, Townsville, 1997, p. 10.
5. Ibid., p. 47.
6. Cheryl Foster, "The narrative and the ambient in environmental aesthetics," *The Journal of Aesthetics and Art Criticism*, vol. 56, 1998 pp. 127–137.
7. Ibid., p. 133.
8. Holmes Rolston, *Conserving Natural Value*, Columbia University Press, New York, 1994, p. 58.
9. Jamie Oliver, "Development of indicators for state of the reef reporting," conference paper, Australian Coral Reef Society, Port Douglas, 17 October 1998.
10. Lucas et al., *The Outstanding Universal Value of the Great Barrier Reef World Heritage Area*, p. 40.
11. R. A. Kenchington, cited in Lucas et al., *The Outstanding Universal Value of the Great Barrier Reef World Heritage Area*, p. 38.
12. Charles Birkeland (ed.), *Life and Death of Coral Reefs*, Chapman & Hall, New York, 1997, p. 11.
13. After Holmes Rolston, *Conserving Natural Value*, p. 73.
14. *http://www2.abc.net.au/science/coral*, forum comment by Ove Hoegh-Guldberg on "How long have corals got?"
15. Global Coral Reef Alliance website, *http://www.fas.harvard.edu/~goreau/restr.accret.html*

16. Terry Done, "Science for management of the Great Barrier Reef," Australian Institute of Marine Science, Townsville, 1998, at *http://www.aims.gov.au/pages/research/smgbr/smgbr02.html*
17. Birkeland, *Life and Death of Coral Reefs*, p. 10.

## SALTWATER SPIRITUALITY

1. David E. Cooper, "Aestheticism and environmentalism," in David E. Cooper and Joy A. Palmer (eds.), *Spirit of the Environment: Religion, Value and Environmental Concern*, Routledge, London, 1998, p. 100.
2. Victor Turner and Edith Turner, *Image and Pilgrimage in Christian Culture*, Columbia University Press, New York, 1979, p. 254.
3. John E. Carroll et al. (eds.), *The Greening of Faith: God, the Environment and the Good Life*, University of New England Press, Hanover, 1997.
4. Miroslav Holub, "Jumping to conclusions," *Island*, vol. 53, 1992, pp. 16–21.
5. Mark O'Connor and Neville Coleman, "The Beginning," *Poetry in Pictures: The Great Barrier Reef*, Hale & Iremonger, Sydney, 1985, p. 10.
6. Dermot Smyth, *Understanding Country: The Importance of Land and Sea in Aboriginal and Torres Strait Islander Societies*, Australian Government Publishing Service, Canberra, 1994, p. 2.
7. Dermot Smyth, "Aboriginal maritime culture in the far northern section of the Great Barrier Reef Marine Park," *Report to the Great Barrier Reef Marine Park Authority*, Townsville, 1991, p. 6.
8. Bruce Rigsby and Athol Chase, "The Sandbeach People and dugong hunters of Eastern Cape York Peninsula: Property in land and sea country," in Nicolas Peterson and Bruce Rigsby (eds.), *Customary Marine Tenure in Australia*, Oceania Monograph 48, University of Sydney, Sydney, 1998, p. 200.
9. As told to Bruce Rigsby and Athol Chase.
10. Dermot Smyth, "Aboriginal maritime culture in the far northern section of the Great Barrier Reef Marine Park," p. 33.
11. A. K. Chase, "Lazarus at Australia's gateway: The Christian mission enterprise in Eastern Cape York Peninsula," in Tony Swain and Deborah Bird Rose (eds.), *Aboriginal Australians and Christian Missions: Ethnographic and Historical Studies*, The Australian Association for the Study of Religions, Bedford Park, S.A., 1988, pp. 127–139.
12. Bruce Rigsby and Athol Chase, "The Sandbeach People and dugong hunters of eastern Cape York Peninsula: Property in land and sea country," in *Customary Marine Tenure in Australia*, p. 198.

13. Patrick Dodson, "The land our mother, the church our mother," *Compass Theology Review*, vol. 22, 1988, p. 1.
14. Robert Johannes, *Words of the Lagoon: Fishing and Marine Lore in the Palau District of Micronesia*, University of California Press, Berkeley, 1981, p. x.
15. Andrew Smith, *Usage of Marine Resources by Aboriginal Communities on the East Coast of Cape York Peninsula*, Great Barrier Reef Marine Park Authority, Townsville, 1989, p. 92.
16. Interview with Andrew Smith, "Traditional and contemporary conservation meet in the republic of Palau," *Reefnet*, issue 7, 1998, at *http://www.reefnet.org/issue7/ngo7.html*
17. Rosemary Hill, "On the forest path," *Times Literary Supplement*, 7 April 1995, pp. 3–4.

# Further Reading


## Introduction

Peter Brown, *Augustine of Hippo* (University of California, Berkeley, 1969).

## Going with the Flow

Sarah Allan, *The Way of Water and Sprouts of Virtue* (State University of New York, Albany, 1997).

Connie Barlow, *Green Space, Green Time: The Way of Science* (Copernicus, New York, 1997).

Charles Birkeland (ed.), *Life and Death of Coral Reefs* (Chapman and Hall, New York, 1997).

Daniel B. Botkin, *Discordant Harmonies: A New Ecology for the Twenty-first Century* (Oxford University, New York, 1990).

Mihaly Csikszentmihalyi, *Beyond Boredom and Anxiety: The Experience of Play in Work and Games* (Jossey-Bass, San Francisco, 1975).

Lao-tzu, *Tao Te Ching: A Book About the Way and the Power of the Way*, a new English version by Ursula K. Le Guin in collaboration with J. P. Seaton (Shambhala, Boston and London, 1997).

John A. Long, *The Rise of Fishes: 500 Million Years of Evolution* (University of New South Wales, Sydney, 1995).

Frank N. McGill (ed.), *Masterpieces of World Philosophy* (HarperCollins, New York, 1990).

John Mulvaney and Johan Kamminga, *Prehistory of Australia* (Allen & Unwin, Sydney, 1999).

Charles Sprawson, *Haunts of the Black Masseur: The Swimmer as Hero* (Vintage, London, 1993).

Colin W. Stearn and Robert L. Carroll, *Paleontology: The Record of Life* (Wiley, New York, 1989).

## WHEN THE REEF WAS OURS

Arthur C. Clarke, *The Coast of Coral* (Shakespeare Head, London, 1956).
Thomas Lowah, *Eded Mer: My Life* (Rams Skull, Kuranda, 1988).
Owen Mass, *Dangerous Waters* (Rigby, Adelaide, 1975).

## REEFSCAPE WITH SEA SERPENTS

Bernard Heuvelmans, *In the Wake of the Sea-Serpents* (Rupert Hart-Davis, London, 1968).
Ove Hoegh-Guldberg, *Climate Change, Coral Bleaching and the Future of the World's Coral Reefs* (Greenpeace, 1999).
Henry Reynolds, *With the White People: The Crucial Role of Aborigines in the Exploration and Development of Australia* (Penguin, Ringwood, 1990).
William Saville-Kent, *The Great Barrier Reef of Australia: Its Products and Potentialities*, 1893, facsimile reprint (John Curry O'Neil, Melbourne, 1972).

## THE TEARS OF THE TURTLE

Noel Monkman, *Quest of the Curly Tailed Horses: An Autobiography* (Angus & Robertson, Sydney, 1962).

## SEA GRASS HARVEST

*The Harper Dictionary of Modern Thought* (Harper & Row, New York, 1988).
Daniel Pauly, *On the Sex of Fish and the Gender of Scientists* (Chapman & Hall, London, 1994).

## CHICK CITY

Tom Griffiths, *Hunters and Collectors: The Antiquarian Imagination in Australia* (Cambridge, Melbourne, 1996).

## AN ISLAND IN TIME

Norman Bartlett, *The Pearl Seekers* (Andrew Melrose, London, 1954).
Josephine Flood, *The Riches of Ancient Australia: A Journey into Pre-history* (University of Queensland, St. Lucia, 2000).

## STRESSED-OUT REEFSCAPE

Jan Sapp, *What Is Natural? Coral Reef Crisis* (Oxford University, New York and Oxford, 1999).

## THE BEGINNING OF THE END OF IT ALL

Holmes Rolston, *Conserving Natural Value* (Columbia University, New York, 1994).

# Interesting Web Pages

Bellaqua Personal Submersibles, at *http://www.flinet.com/~gulfstream/bell.html*

Terry Done, "Science for management of the Great Barrier Reef," Australian Institute of Marine Science, 1998, at *http://www.aims.gov.au/pages/research/smgbr/smgbr05.html*

Terry Done, *Science for Management of the Great Barrier Reef*, Australian Institute of Marine Science, 1999, at *http://www.aims.gov.au/pages/research/smgbr/smgbr06.html*

Bob Friel, "The virtual round-table: Industry leaders share their vision of how, where and why you'll be diving in the next millennium," *Scubadiving*, Nov–Dec 1997, at *http://www.scubadiving.com/feature/specials/roundtable/*

Ove Hoegh-Guldberg, *Climate Change, Coral Bleaching and the Future of the World's Coral Reefs*, Greenpeace, 1999, at *http://www.greenpeace.org.au/info/archives/climate/index.html*

"Farming the sea," interview with Peter Rothlisberg, CSIRO Aquaculture and Biotechnology Program, by Sally Dakis, ABC Ocean Week Feature, at *http://www.abc.net.au/rural/oceanweek/feature2.htm*

The Global Coral Reef Alliance website at *http://www.fas.harvard.edu/~goreau/restr.accret.html*

Ove Hoegh-Guldberg, *Climate Change, Coral Bleaching and the Future of the World's Coral Reefs*, Greenpeace Australia, 1999, at *http://www.greenpeace.org.au/info/archives/climate/index.html*

*http://www2.abc.net.au/science/coral*, forum comment by Ove Hoegh-Guldberg on "How long have corals got?"

K. Hulsman, P. O'Neill, T. Stokes, and M. Warnett, "Threats, status, trends and management of seabirds on the Great Barrier Reef," *The Great Barrier Reef: Science, Use and Management*, 1996, at *http://www.reef.crc.org.au/3reports/conference1/volume1/s3_4.html*

"Marine bioprospecting for the National Cancer Institute," *Reef Research*, issue 7, 1998, at *http://www.reefnet.org/issue7/research7.html*

*http://www.nationalgeographic.com/resources/ngo/education/xpeditions/resources.html*

The NOAA website is at *http://www.pmel.noaa.gov/toga-tao/el-nino/nino-home.html*

Michelle Richards, "Options for introducing BRD's (bycatch reduction devices) onto prawn trawlers in the Great Barrier Reef Marine Park," at *http://cathar.tesag.jcu.edu.au/~tgpjc/TG3201/Background/BRDs/brd.htm*

Interview with Andrew Smith, "Traditional and contemporary conservation meet in the republic of Palau," *Reefnet*, issue 7, 1998, at *http://www.reefnet.org/issue7/ngo7.html*

"Sunfish NQ response to Queensland east coast trawl fishery: draft management plan, 4 July 1999," at *http://www.sunfish.org.au/news/Trawl_response.htm*

"Sustainably managing Queensland's dwarf minke whale tourism industry," CRC Reef Research Center Newsheet, James Cook University, May 1999, at *http://www.gbrmpa.gov.au/~crcreef/4news/News/news202.html*

Barry Tobin, "How the Great Barrier Reef was formed," Australian Institute of Marine Science, 1998, at *http://www.aims.gov.au/pages/research/project-net/reefs/apnet-reefs03.html*

Amanda Vincent, "Marine species in traditional medicine," *Reef Encounter*, vol. 23, 1998, at *http://www.uncwil.edu/isrs/reef-encounter/re23/reef_encounter23.htm*

The Women's Scuba Association, at *http://www.womeninscuba.com/wsa/history/index.html*

# Acknowledgments

There are many people I must thank for help they provided both directly and indirectly. Scientists at James Cook University were most generous with their time. John McLaren asked me to write about the Chesterfield Islands when he was editor of *Overland*, and this book has grown from an article he published. Stephanie Holt, editor of *Meanjin*, gave encouragement at one of those black moments of despair when the writer is convinced the right words will never come.

Archives of early films of the reef, many in sepia, or black and white, or in colors now badly faded after 60 years in the can, proved immensely valuable in showing the reef as it first appeared to the world in newsreels and films. I was able to view these films in the collections of the National Film Archive, Canberra and Cinemedia, Melbourne.

I would also like to thank the following: Ameer Abdulla, Max Allen, Paul Armsworth, Jim Charley, Brooke Carson-Ewart, Elizabeth Cameron, Howard Choat, Barbara Erskine, Maurita Harney, Hal Heatwole, Joanna Hugues, Sylvia Kelso, John A. Long, Harold Love, Brian McFarlane, Helene Marsh, Richard Overell and the Rare Books Room of Monash Library, Jan Patterson, Peter

Pierce, Len Radic, Therese Radic, Rob Riel, Nonie Sharp, Barbara Waterhouse, and especially Jackie Yowell of Allen & Unwin. I am grateful to my fellow diving students, Vickie, Deb, Charlene, Vindie, Tom, Stephie, and John, and instructor Jason, for keeping me both afloat and safely submerged.

I would like to thank the many excellent librarians who helped me, from the Australian Museum in Sydney, the State Library of New South Wales, the State Library of Victoria, James Cook University, Monash University, and the Great Barrier Reef Marine Park Authority.

I have been very fortunate indeed to have made many reef trips in the company of members of the science departments of the University of New England, Armidale, New South Wales, Australia. My sister Kathleen King, now of the University of New England, introduced me to the UNE reef expeditions, and without her input this book would never have started, nor would we have enjoyed so many brilliant trips together. Hal Heatwole and Jim Charley provided the scientific leadership of various expeditions, always with Max Allen as skipper, of first the *Australiana*, then the *Kanimbla*, and later *The Spirit of Freedom*. I am grateful to Max and Carmel Allen for taking me to so many special places on the reef, in luxury and safety. Someone (not I) should sit down with Max and his guitar and record his songs of early Queensland, both printable and unprintable, before they are lost.

I completed this book while a senior research associate in the School of Literary, Visual and Performance Studies, Monash University, Melbourne, and while a visitor to the English Department, James Cook University, and John Flynn College, Townsville.

Earlier versions of some material printed here have been published in *AQ*, *Arena Magazine*, *Meanjin*, *Overland*, and *Paradoxa*.

# Index

## B

## D

## G

## H

## I



## J

K



L



M

## U



## V



## W

Y



Z